THE WORLD WE'LL LEAVE BEHIND

It is now clear that human activity has influenced how the biosphere supports life on Earth, and given rise to a set of connected environmental and social problems. In response to the challenge that these problems present, a series of international conferences and summits led to discussions of sustainable development and the core dilemma of our time: How can we all live well, now and in the future, without compromising the ability of the planet to enable us all to live well?

This book identifies the main issues and challenges we now face; it explains the ideas that underpin them and their interconnection, and discusses a range of strategies through which they might be addressed and possibly resolved. These cover things that governments might do, what businesses and large organisations can contribute, and the scope for individuals, families and communities to get involved. This book is for everyone who cares about such challenges, and wants to know more about them.

William Scott is Emeritus Professor of Education at the University of Bath, UK, and is Chair of Trustees of the UK's National Association for Environmental Education.

Paul Vare is Postgraduate Research Lead for the School of Education at the University of Gloucestershire, UK, and a Director of the South West Learning for Sustainability Coalition.

"What a wonderfully readable cornucopia of information and ideas – an inspiring and practical source for confidently talking about and acting on behalf of a sustainable future!"

— Sara Parkin, The Sustainability Literacy Project

"This is a lucid collection of short chapters on some of the world's most intractable issues like globalisation, climate change, human population and migration. It is a unique and prescient contribution that helps make sense of an increasingly unpredictable and unsustainable world."

— Professor Stephen Martin, Honorary Professor,
University of Worcester, and Visiting Professor in
Learning for Sustainability, University of the West of England, UK

"These informed authors help readers understand vital issues about sustainability and how we live today. Importantly, they set out a range of strategies whereby everyone who's interested can play a role in keeping our planet living for generations to come."

— Per Sund, Associate Professor, Stockholm University, and
Research Fellow, Rose Project, Karlstad University, Sweden

"I loved this book. The authors have written short but informative, provocative, and amazingly readable chapters on a wide range of issues, concepts and strategies that help us think about how we might create a more sustainable and just future. The format is brilliant!"

— Judy Braus, Executive Director,
North American Association for Environmental Education

"Scott and Vare have written a very clear, accessible introduction to the sustainability idea and some of its difficulties – an excellent book for people coming new to the topic, and one from which everyone can gain new insights."

— John Foster, Lancaster University, UK

THE WORLD WE'LL LEAVE BEHIND

Grasping the Sustainability Challenge

William Scott and Paul Vare

Routledge
Taylor & Francis Group

LONDON AND NEW YORK

First published 2018
by Routledge
2 Park Square, Milton Park, Abingdon, Oxon OX14 4RN

and by Routledge
711 Third Avenue, New York, NY 10017

Routledge is an imprint of the Taylor & Francis Group, an informa business

© 2018 William Scott and Paul Vare

British Library Cataloguing-in-Publication Data
A catalogue record for this book is available from the British Library

Library of Congress Cataloging-in-Publication Data
A catalog record for this book has been requested

ISBN: 978-1-78353-774-7 (hbk)
ISBN: 978-1-78353-773-0 (pbk)
ISBN: 978-1-351-24293-6 (ebk)

Typeset in Joanna
by Swales & Willis Ltd, Exeter, Devon, UK

MIX
Paper from
responsible sources
FSC
www.fsc.org   FSC® C013604

Printed and bound by CPI Group (UK) Ltd, Croydon, CR0 4YY

This book is dedicated to children and grandchildren across the planet, especially Lilly, Levi, Ellie, Bathsheba, Asha and Amber.

CONTENTS

PART II
Concepts   **91**

FIGURES

TABLES

INTRODUCTION

The thrushes sing as the sun is going,
And the finches whistle in ones and pairs,
And as it gets dark loud nightingales
In bushes
Pipe, as they can when April wears,
As if all Time were theirs.

These are brand new birds of twelvemonths' growing,
Which a year ago, or less than twain,
No finches were, nor nightingales,
Nor thrushes,
But only particles of grain,
And earth, and air, and rain.

Thomas Hardy, "Proud Songsters"

Our purpose

We have written this book for anyone who has an interest in the environmental and social challenges facing humanity today. If you would like to have a brief introduction to these challenges, the ideas that help explain them and some of the possible strategies for addressing them, then this book is for you. We do not set out to provide an in-depth account of the problems facing us, or a deep historical perspective on their development. Neither is this a toolkit, self-help book or a set of instructions for good living. Books about environmental and social sustainability with these purposes already exist. Rather, we set out to summarise ideas in a way that will help you join in the public and political debate, and help you think about the state of the world that we shall all leave behind.

The sustainability challenge

All life on Earth depends on the Earth for its continued existence. As Thomas Hardy's poem reminds us, the food we eat, the water we drink, the raw materials we use to make the stuff we buy and sell, all come from the biosphere, that thin layer of soil, rock, water and air around our planet where all life is found. By contrast, all the energy we use ultimately comes from the Sun.

Even though we cannot control the weather, ocean currents or the great cycles of nature, or predict earthquakes, we now have a sophisticated understanding of how all these work. We are also able to know, as never before, when it all goes wrong: images of raging forest fires, extreme droughts, extensive oil spills, vast floods, the continuing loss of valuable land, and people in distress appear on our screens on a daily basis – as do heroic attempts to stem this tide of trouble.

It is now clear that human activity has influenced how the biosphere supports life on Earth. Examples of this include the loss of unique habitats and species, the acidification of the oceans, air and water pollution, and now rapid climate change. As many of these issues worsen, it's now widely accepted that the huge inequalities between people across the world are a source of instability as well as an affront to human dignity.

Yet, all is not lost. Indeed, we wouldn't have written this book if we thought it were. In response to the challenge that these issues present, there have been a series of international summits and agreements that have led to the idea of sustainable development. This term, although tricky to define, covers the attempt to meet the core challenge of our time: How can we live well, now and in the future, without compromising the planet's ability to enable us all to live well?

The responsibility to do something applies with some force to governments, business and organisations of all kinds, particularly in relation to rapid climate change and rising demand for raw materials and energy. But it also applies to communities and families. Whether we think of the cost of energy and food, the effect of a changing climate on coasts and rivers, the provision of overseas aid and help to refugees, or a wide range of other priorities, there is a need for us all to be more aware of these issues, the ideas that underpin them, and proposed solutions to them. To make all this clearer, and to explore the issues, are the reasons we have written this book.

Structure

Our 55 self-contained chapters are all short, typically between 900 and 1,100 words. Each one focuses on a key idea in what we hope you will think is an open-minded, clear and valid way, setting out the essence of the idea while also showing its connections to others. These ideas are presented in three parts: issues, concepts and strategies, although the distinction between these is not always as clear-cut as this division suggests. This is because one thing is inevitably connected to another; usually to many others.

Part I: Issues

These are the real challenges we now face because of our own past and present activities.

Many of these are problems we have created in the biosphere by how we have acted. These include: warming the atmosphere and ocean, depleting the ozone in the stratosphere, making the seas more acidic, polluting the air and sources of fresh water, destroying habitats and ecosystems, reducing biodiversity and driving species to near extinction, and, of course, rapid climate change. They also include the consequences of these actions, such as food scarcity and famine, and the plight of refugees. But some of the challenges are different. These are the problems that we have created by how we think about other people, for example inequality, gender disparity, discrimination, and the lack of environmental justice, which are, in turn, related to issues of economic growth and the global human population.

Part II: Concepts

These are the ideas and mental frameworks that we use us to think about and understand these challenges.

Some of these are about understanding nature and the biosphere, ideas such as biodiversity and phenomena such as the greenhouse effect, and the idea of the Earth as a living organism: Gaia. Some are concerned with how we think about the relationships between humanity and the biosphere; for example, the Anthropocene, systems thinking, harmony, complexity and resilience. Others are about the ideas we have developed to explain what has gone wrong, and what we might do to improve matters, such as sustainable development and its 17 global goals, and how such ideas are framed. Some are about notions such as globalisation and neoliberalism that concern the politics of global commerce.

Part III: Strategies

These are the means through which such challenges might be addressed and possibly resolved or overcome.

Some of these are things that governments can do, through policies, legislation and formal education programmes. Some are social, such as the transition movement, and are led by ordinary people, and some arise from international agencies and global charities – for example, the Sustainable Development Goals and the Earth Charter, the protection of endangered species, and the identification of protected areas for wildlife. Some are what business can initiate – for example, the circular economy and carbon capture and storage. Alongside these practical strategies, there are also particular ways of thinking and approaching problems, for example through biomimicry and rewilding.

There are notes to each chapter that provide the full titles of books we have mentioned, and links to the websites of organisations whose work we have discussed. There are also web links to reports, discussion documents, articles, data sets, blogs, etc. Where web addresses are long, we have tended to use the shortened ow.ly/xxx format. Copying these into a web browser should take you directly to the correct web page.

Reading the book

We think there are several ways that you might read this book. Obviously, starting with Chapter 1 and reading through to the end is one way, although you would be hard-pressed to find a single storyline. So we also recommend dipping in as interest takes you, moving from one focus to another.

Although we have put the issues first, because that makes sense to us, we know that some readers will find concepts or strategies of greater interest, and so you might start with Chapter 21 or Chapter 39. It is also quite possible to jump from part to part and back again in order to follow a theme, for example:

Natural resource depletion → Recycling → Frames and framing → The circular economy → Biomimicry

Inequality → Gender disparity → Development → Reducing global inequalities → Feeding 10 billion

Species loss → Biodiversity → Elephants, rhinos and donkeys → Protected areas → Rewilding

Global warming and climate change → The greenhouse effect → Carbon capture and storage

Environmental justice → Globalisation → Neoliberalism → Brexit and environmental law

Ozone and CFCs → Systems and systems thinking → Complexity → The Montreal Protocol

Economic growth → Sustainable development → Neoliberalism → Living within limits

Biocentrism → Gaia → Harmony → Living within limits → The Earth Charter → (Environmental) education

Try it and you'll see what we mean. Whichever route you take, we hope to have piqued your interest enough so that you feel inspired to follow things up yourself. Finally, although there is no conclusion chapter in the book, there are three questions we'd like to leave you with:

What do I think about all this?

Where does it leave me?

What might I now do?

We ask these questions all the time, and they summarise the reasons why we wrote the book. We hope they are useful to you too.

PART I

ISSUES

1

GLOBAL WARMING AND CLIMATE CHANGE

The Earth is warmer than it ought to be, given its distance from the Sun, because some of the gases in the atmosphere create what is called a greenhouse effect. This is a natural phenomenon, although it has little to do with how a greenhouse actually works. It is also a good thing as, without it, the Earth would be so cold as to make life as we know it very difficult. The extra warming is caused by so-called greenhouse gases as more heat is retained in the atmosphere than otherwise would be. The main greenhouse gases are carbon dioxide, methane, nitrous oxide and water.

Since the start of the Industrial Revolution, the amount of some of these gases in the atmosphere has been increasing, particularly carbon dioxide (CO_2), and so has the warming. Changes in CO_2 concentrations over the past 650,000 years have been measured by examining the composition of air trapped in Antarctic ice.

This is the current picture of a warming Earth:

- From 1880 to 2012, average combined land and ocean surface temperature rose by $0.85 \pm 0.2°C$. Early in 2017, NASA said that the increase had become $1.1°C$.[1]

- Ocean warming is largest near the surface, and the upper 75m has warmed by 0.11 ± 0.2°C per decade since 1970.
- Since 1750, the uptake of CO_2 has resulted in a 26% increase in the acidity of the oceans, which is bad for coral reefs and fish.
- From 1900 to 2010, global mean sea level rose by 0.19 ± 0.02m. This is larger than the rate during the previous 2,000 years.
- From 1980 to 2012, the area covered by Arctic sea ice has decreased, by about 4% per decade, leading to more absorption of solar radiation.

What about the future?

Greenhouse gas emissions are mainly driven by population size, economic activity, lifestyle, energy use, land use patterns, technology and climate policy, and there is a consistent, almost linear, relationship between cumulative CO_2 emissions and projected global temperature change up to the year 2100. While a warming Earth is a problem in itself, it's the effect on climate that is the real concern.

How much our climate changes in the future, and how it changes, depends on how much warmer it is going to get. In turn, this depends on what happens to the amount of greenhouse gases in the atmosphere. In terms of CO_2, which contributes about 70% to warming, this depends on how successful we are in reducing the CO_2 emissions that the global economy generates.

The UN's Intergovernmental Panel on Climate Change (IPCC) has set out four scenarios that describe different pathways for future greenhouse gas emissions.[2] There is a very tight scenario (where emissions are kept low), two intermediate ones, and a loose scenario (where emissions continue to be high). The first of these assumes global warming to be no more than 2°C above 1750 levels. In 2015, it was 0.85°C above that level, and is now higher.

Future climate will depend on past and future emissions, and on natural climate variability. The IPCC says that the surface temperature change from 2016 to 2035 (compared with 1986 to 2005) is likely be between 0.3°C and 0.7°C, assuming no major volcanic eruptions, no changes in natural sources of methane and nitrous oxide, or any unexpected changes in total solar radiation. It says that warming will continue beyond 2100 under three of its scenarios and that surface temperatures will stay at elevated levels for many centuries, even when the economy stops creating CO_2.

The IPCC says that other changes could last thousands of years after the global surface temperature is stabilised, and that a reduction in the amount of permafrost is virtually certain if global temperatures continue to rise. This could result in a large release of methane – another strong greenhouse gas – which is trapped within the permafrost, and would likely cause another rise in temperatures, and result in feedback loops leading to even more warming.

The IPCC says that many species will face increased extinction risk during and beyond the twenty-first century. This is because most plant species cannot shift their geographical ranges sufficiently fast to keep up with how quickly climate changes. Most small mammals and freshwater molluscs will also face this problem. Marine organisms will face progressively lower oxygen levels and high rates of acidification within the oceans, affecting the vitally important phytoplankton. These carry out photosynthesis, and are important food stocks for fish and whales. Coral reefs and polar ecosystems are highly vulnerable, and coastal systems and low-lying areas are at risk from sea level rise, which will continue for centuries.

Climate change will also likely reduce food security, particularly the availability of staple crops such as wheat, rice and maize. To begin with, it will exacerbate existing human health problems, but will then probably lead to increases in ill health in many regions, and especially in countries with low incomes. It will probably increase human migration on a huge scale, and may indirectly increase the risks of violent conflicts by increasing poverty and economic shocks.

Of course, none of this has to happen; the IPCC is just saying that it might if governments do not act to reduce CO_2 and other greenhouse gas emissions. A UN conference in Paris in December 2015 attempted to reach agreement on doing something about all this, and resulted in the Paris Agreement, which set a new target to limit global temperature rises to no more than 1.5°C above pre-industrial levels. These are some of the key points of the agreement:

- This marks a global political recognition of the reality and risks of climate change.
- On the positive side, there is an ambitious goal of having as much greenhouse gas coming out of the atmosphere as going into it in the second half of this century.

- On the negative side, existing national pledges on climate action are more in line with a warming of 3°C than 1.5°C.
- The agreement requires $100 billion a year to move from economically developed countries to developing ones by 2020, with the amount to be reviewed in 2025. This obligation also bears (morally, at least) on rich developing countries as well: India, China, etc. It remains to be seen how well all that money will be spent.

The Paris Agreement requires countries to act, and most have now signed up to it.[3,4,5] However, it says nothing concrete about what or how much they have to do; all that will need to be negotiated. The agreement marks a transition away from the age of fossil fuels, although this may well be more drawn out than many would wish, or that society needs to live well. From now on, however, the idea of investing in coal and oil may seem riskier than it once was.

Notes

1 The NASA website has a wide range of data on climate change and global warming, and excellent charts and pictures. Web link: climate.nasa.gov.
2 The IPCC website has data on climate change and details of its methodology. Web link: ipcc.ch.
3 The Paris Agreement website has a number of reading between the lines comments on the agreement and its implementation. Web link: parisagreement.org.
4 The UN Climate Change Newsroom has a range of articles and agreements about climate change, including the Paris Agreement. Web link: unfccc.int.
5 In June 2017, President Trump announced that the US would be withdrawing from the agreement in November 2020 – the earliest possible withdrawal date.

2

INEQUALITY

In early 2017, *The Times* columnist Philip Collins wrote: "We are living through a long arc of progress."[1] He meant by this that democracy was thriving, with more than 4 billion people now living under a democratic government; that is, one that can be removed from power through open and fair elections. Collins saw this as a good thing because, as he put it, "Democracies are hosts to free lives."

He cited a number of positive developments, which included:

- all Western countries recognise same-sex unions;
- the pan-African parliament endorsed a continent-wide ban on female genital mutilation;
- the number of people living in extreme poverty dropped below 10% for the first time; and
- maternal mortality has fallen 44% since 1990, with infant mortality halving over the same period.

After listing more of such good news items, many of which are explored in this book, Collins said it was no wonder that "Europeans are, on average, at their happiest since the [2008] financial crisis." But averages are deceptive things, which, by their nature, tend to hide more than they reveal. And Europeans, whether happy or unhappy, are not the only people on the planet, even if they may be perceived as being among the luckier ones because of their comparative wealth and material advantages.

There are, of course, poor and rich families in every country, no matter how economically (under)developed it is. Because of this, it is necessary when thinking about inequality to consider differences in wealth and income within countries, as well as between them. In his recent book, François Bourguignon says this about in-country inequality: "After a significant decline in the mid-twentieth century, followed by a long period of stability, inequality has begun to rise over the last two or three decades in the large majority of developed countries."[2] This rise in income inequality has not been uniform, however. While the US and UK have seen exceptionally large rises in income inequality over the past 35 to 40 years, there have been smaller rises in Canada, Germany and the Netherlands. France, meanwhile, has seen a slight reduction. In terms of the absolute levels of inequality across the EU, the Nordic countries have the lowest and the UK and Italy the highest, with others in between. All this suggests that domestic policies, for example about tax, welfare and investment, may be in play, and that the owners of capital assets such as property, shares and pensions have seen their overall wealth rise faster than people without such things.

Bourguignon says that income inequality has also risen within a number of developing countries for which there are long-term data. These include Bangladesh, China, India and Indonesia, although not other Asian countries. Many African countries, for example Ghana, Kenya and Nigeria, have also seen such large inequality increases, but most have not, while in Latin America, inequality remains high.

However, while inequality has been rising within such countries, it has fallen at the global level over the last 20 years. It has done so from levels that were very high because of different rates of economic development over the past 200 years. The most significant contribution to this fall has been the recent rapid industrialisation in China, and to a lesser extent in India. These two countries, between them, have 40% of the global population. This development has resulted in a reduction in the proportion of the world's population living in extreme poverty, as these statistics illustrate: in 1990, 47% of the population in the developing world lived on less than $1.25 a day; in 2015, it was 14%. Globally, the number of people living in extreme poverty fell from 1,900 million (in 1990) to 836 million (in 2015).[3]

Globalisation and neoliberal economic policies have played their part in these changes, as has technology, and it is notable that four of the eight richest men in Oxfam's 2017 survey of global wealth were in the IT/data business.[4] Oxfam's report made the point that countless billions of people have no financial wealth at all, other than their very modest incomes. Earnings data also illustrate this disparity very clearly. The richest 600 million people on Earth have an average income of $25,000 per annum, with the poorest 600 million having an average of $270. This is a ratio of 90:1. In Brazil, François Bourguignon says the ratio is (only) about half that, but it is still 50:1.

There are many contributory factors to the size of these ratios and to why they have been increasing. These include vast (and to many people completely unwarranted) earnings in the banking, financial and business sectors where there sometimes seems little effective control of either earnings or corporate behaviour. In many countries, there seems to be much less animus against those working in media, culture and sport who have large incomes than there is against those in the corporate sector, even when actors, singers, authors, celebrities and footballers are seen to use aggressive schemes to avoid paying tax. Whether this is because the contribution of such people to our wellbeing and happiness is more valued than, say, a banker's, or because we know we have a choice about whose music we download or whose books we buy, is an open question.

An important issue in relation to inequality is: How much does it matter? After all, in a developed market economy, income equality would be impossible to maintain, and the prospect of greater earnings can be motivational. This raises the question as to whether there are income ratios above which inequality becomes a problem with serious consequences for societies and people within them. In their book *The Spirit Level*, Richard Wilkinson and Kate Pickett present evidence to demonstrate the ill effects of high inequality levels.[5] These range from a lack of trust, to anti-social behaviour, to a host of mental health issues. They also show that after a certain modest level of individual wealth is achieved, key indicators of wellbeing, such as life expectancy and a sense of happiness, no longer increase as societies grow richer.

Looking at inequality within countries, it is obvious that all countries struggle with finding the most appropriate balance between generating wealth and redistributing it for the greater social good; that is, between economic effectiveness and efficiency, on the one hand, and fairness and social cohesion, on the other; between incentivising people to work hard and do well, while encouraging them to willingly contribute to the social commons such as education, health, welfare and security. Social and educational investment in the young seems a sensible public policy no matter where you live, especially if it can be targeted towards those born into social disadvantage.

Viewing inequality between countries is more difficult from a public policy perspective. A contribution to what is usually still called overseas aid remains important, even if it does not quite reach the 0.7% of national income that the UN suggests (and that the UK manages). That such aid needs careful targeting seems obvious, both to help people directly and to keep the cash out of the hands of well-paid consultants, corrupt middlemen and venal government ministers. This is a challenge for us all.[6,7]

Notes

1 Collins, P. (2016). Never forget that we live in the best of times. *The Times*, 23 December.

2 Bourguignon, F. (2015). *The Globalisation of Inequality*, translated by T. Scott-Railton. Princeton, NJ: Princeton University Press.

3 The UK Office of National Statistics (ONS) has up-to-date and historical data on inequality in the UK. The OECD carries similar data for economically developed countries, and the World Health Organisation (WHO) has data on global health inequality. Web links: ons.gov.uk, oecd.org, and who.int.

4 Oxfam issued a press release on 16 January 2017 saying that "Just 8 men own same wealth as half the world." Web link: oxfam.org.uk. In a press release on 18 January 2017, Oxfam was criticised by the Adam Smith Institute for its approach to these data and the assumptions it made. Web link: adamsmith.org.

5 Wilkinson, R. & Pickett, K. (2010). *The Spirit Level*. Harmondsworth: Penguin.

6 Atkinson, A.B. (2015). *Inequality: What Can Be Done?* Cambridge, MA: Harvard University Press.

7 The Equality Trust is a charity that works to improve the quality of life in the UK by reducing economic inequality. Web link: equalitytrust.org.uk.

3

SPECIES LOSS

In June 2015, The Guardian ran this headline: "Humans creating sixth great extinction of animal species, say scientists".[1] The paper was reporting on a study which argued that the rate of extinction for species in the twentieth century has been up to 100 times higher than it would have been without human impact. We are not good for other creatures it seems, despite much religious encouragement over millennia to be stewards of the Earth.

Estimates of the number of separate species on Earth vary, but around 10 million seems near the mark. Somewhere between 1.2 million and 1.5 million of these have actually been recorded, which means that about 85% have not yet been identified. All such numbers come with considerable uncertainty.[2]

Some argue that the natural rate at which species extinction occurs is two species per 10,000 in every 100 years; others say it's only one per 10,000. Assuming 10 million species, this means that every 100 years, between 100 and 200 species vanish from the Earth — most of which we never knew about. If the authors of the report quoted in The Guardian are right, the real number, thanks to human impact, will be between 10,000 and 20,000 species per 100 years. This seems a lot.

Estimates vary, of course, but it is conservatively thought that around 5 billion species have existed at one time or another on the Earth, which means that over 99% of them are now extinct. The point to remember in all this is that species come and go over time, as Charles Darwin explained. It's as natural a process as the Sun shining, an oak tree growing or a volcano erupting.

The Guardian headline said "sixth great extinction," which implies that there have been five previous ones when swathes of species disappeared. Evidence from the fossil record suggests that these were (in millions of years ago):

- 435: 60% of all kinds of both terrestrial and marine life worldwide vanished.
- 360: 75% of all species died out.
- 250: 80–95% of all marine species went extinct.
- 205: About 50% of marine invertebrates and 80% of all land quadrupeds went extinct.
- 65: The end of the dinosaurs. Hardly any large land animals survived; plants and tropical marine life were also much affected. In evolutionary terms, this is quite recent.

All these were natural events, one way or another. The point of the Guardian story is that the sixth extinction is unnatural, that is, all very human influenced. Species loss can happen for a range of reasons, many of which are as a result of human development, including loss of habitat, pollution, poaching and competition from other species. Naturally enough, not everyone agrees that this is now happening on the scale that the word extinction implies, but 2016 saw the first reported loss of a mammal species (the Bramble Cay melomys) because of sea level rise caused by climate change. This small rodent was only found on a very small low-lying island in the eastern Torres Strait off Australia, and so was particularly vulnerable.

But what is a species anyway? And should we be concerned about loss? A species is the lowest level of animal classification. Take birds, and think of the European greenfinch. This is a separate species that has many features in common with five other greenfinch species. These are said to be all part of the same genus, which is thought of as a group of very closely related species. The classification goes:

Table 3.1 The animal classification

Species	Chloris chloris	The European greenfinch
Genus	*Chloris*	Five species of greenfinch
Family	*Fringillidae*	Fifty-two genuses, including all the finches
Order	*Passeriformes*	About 100 families (5,400 species) of birds, including finches, warblers and sparrows
Class	*Aves*	All the birds
Phylum	*Chordata*	All animals with backbones
Kingdom	*Animalia*	All animals

Source: Wikipedia

Wikipedia's introduction to its page on the greenfinch sums all this up as: "The European greenfinch, or just greenfinch (*Chloris chloris*), is a small passerine bird in the finch family Fringillidae."[3]

And is species loss important?[4] It's certainly important to distinguish between the entire loss of a species from the Earth and its disappearance from a particular place. The wolf and lynx no longer roam around the UK because they were hunted until they were no more, but they still thrive elsewhere in the world. It is confusing to say, as some do, that these animals are extinct in the UK; much better, perhaps, if more long-winded, to say that they're no longer found here in the wild.

But is it a problem that there are no longer wolves in the UK? Inevitably, this depends on your point of view. The absence of wolves means that large animals such as deer are no longer killed by other animals and have to be controlled in other ways, such as by human hunting. On the other hand, sheep farmers in the hills shed no tears for the missing wolf as their flocks are safer in our wolf-less, but less natural, times. Many of those who think that rewilding the land is a good idea would like to see the wolf and lynx return, and they tend to think that sheep farming on the hills should stop anyway.

So, does species loss matter? After all, it is the end of the line. "Yes," say those who think all creatures have value in their own right. "No," say others; it's a natural process and we should let it happen. "It depends," say others, on which species goes. "Yes," if it's a so-called keystone species whose loss can have a disproportionately devastating effect on an ecosystem; this, the argument goes, is like removing the keystone from an arch – though ecosystems are much more complex than arches. Such species are said to play a critical role in maintaining the integrity and quality of an ecosystem. But "no" for species at the end of an evolutionary cul-de-sac, such as the giant panda; here, it would probably make little if any actual difference to anything, although some animal charities would need to find another charismatic megafauna to persuade us to donate our money.

There is usually some relativism in all such arguments, as, although there's a clear moral argument against human-induced species loss, not everyone would extend this to the smallpox or Ebola viruses.

Notes

1 Vaughan, A. (2015). Humans creating sixth great extinction of animal species, say scientists. *The Guardian*, 19 June. Web link: ow.ly/6NhK3oaxD9N.
2 IUCN has been assessing the conservation status of species, subspecies, varieties and sometimes selected sub-populations on a global scale for the past 50 years in order to highlight threats of extinction and to promote conservation. The IUCN Red List contains species that are the key building blocks of ecosystems. Web link: iucnredlist.org.
3 Wikipedia on the greenfinch. Web link: en.wikipedia.org/wiki/European_greenfinch.
4 WWF's website is also a source of information on threatened species. Web link: wwf.panda.org.

4

HUMAN POPULATION

People worry about population for all sorts of reasons and at different scales. They worry about it globally:

- so many people in an overcrowded world;
- too many mouths to feed, and not enough food for them; and
- huge numbers of children dying from preventable diseases.

Then there are regional and national issues:

- all those economic migrants and refugees heading their way wanting their jobs;
- migrants with the wrong sort of ethnicity, culture or religion not fitting in; and
- ageing populations with too few young people to replace them.

And some very local ones:

- not enough social infrastructure (schools, hospitals, etc.) to cope with people going to particular places; and
- too much local housebuilding that will attract more people to small towns and villages and change their character.

These are not separate worries, but are connected, sometimes in unexpected ways. This chapter is about the nature of the global human population, which has seen huge increases over the last few hundred years. Past figures were:

Table 4.1 Changes in human population size

Year (CE)	Population (million)
200	190
1000	275
1200	360
1500	450
1700	610
1760	770
1804	1,000
1900	1,600
1927	2,000
1960	3,018
1970	3,682
1980	4,400

Source: UN

According to the UN, in 2015, there were 7,350 million people living in the world, and in 2050 the UN estimates it will be about 9,730 million.[1]

By the time you read this, these numbers will have changed, but if you go to the Worldometers website, you can watch them, and all sorts of other global data, changing in real time.[2]

Although the year-on-year increases now seem to be getting larger, the % growth is actually slowing. It was 1.85% in 1951, 2.09% in 1969, but only 1.13% in 2016. It could be that, if this trend continues, the global population will stop rising, and remain steady or even fall. That said, a 1.1% rate of growth currently adds about 83 million people to the world every year – to eat and drink, to be clothed, to consume, and to have children in their turn. Eighty-three million is just a bit bigger than the populations of Iran and Germany, and if we wanted to scare you, which we don't, we'd probably say that was an enormous number.

The world's population is not evenly distributed, with India (1,326 million) and China (1,382 million) having the largest national numbers. We've put India first as its population growth rate is about three times that of China (and its fertility rate is about twice as high), and so their positions at the top of the population league table will soon switch over. No other country comes anywhere near these two behemoths, each of which has about 18% of the global population. All these are 2016 numbers.

Population growth rates vary widely as well. Among the larger countries, these range from the negative: −0.44% (Ukraine) and −0.2% (Japan) to the positive: +3.23% (the Democratic Republic of the Congo) and +3.155 (Tanzania). Smaller countries show a similar range. The reasons for these vary as well: gender inequality (particularly female education rates), war, economic chaos, disease and pandemics, political instability, high (or low) birth rates, good (or terrible) healthcare, and a readiness (or not) to welcome strangers to the country.

The fertility rate needed to maintain a steady population is 2.1 children per woman. This is the average number of live births a woman would need to have (by age 50) to 'reproduce herself' by bearing a daughter who survives to childbearing age. Its calculation assumes that there is no mortality; this is never the case in reality, even in countries with the very best healthcare systems.

There are now many countries across the world where the fertility rate is either well above or below this.

Here are some contrasting fertility rates:

Table 4.2 Examples of fertility rates across the world

Somalia	6.61
Mali	6.35
DR Congo	6.15
Burundi	6.08
Bangladesh	2.12
USA	1.89
Japan	1.40
Singapore	1.23

Source: UN

Low fertility rates can result in an ageing population, with fears that a country's economy will suffer as a smaller and smaller proportion of the population is available for work. There are several ways around this:

- welcoming migrants;
- bringing in guest workers (who usually have few rights, don't get paid much, and are expected to go home at some point);
- economic incentives to encourage childbirth and/or social disincentives to reduce its avoidance through, say, abortion;
- raising the official retirement age; and
- investing heavily in technology to replace people in the workplace – automation, computing and robots.

Those countries whose cultures are not welcoming of migrants, and/or whose languages are tricky to master, are most vulnerable to these difficulties. Some think that Germany's official welcoming of Syrian refugees in 2015 had at least something to do with overcoming its low fertility rate. By contrast, the UN says high fertility rates are associated with increased risk of serious maternal illness and death, as well as high infant mortality.[3] These tend to be found in poorer countries where social and health services are less developed.

The next table shows UN mortality rates per 1,000 live births in 2015 for under-fives.

It is data such as these that have encouraged the world to adopt the 17 Sustainable Development Goals (SDGs) in order to try to narrow (and perhaps even eliminate one day) such huge discrepancies. Doing something to prevent and adapt to a changing climate is at the heart of the goals, as a changed climate could make this task much harder than it already is, because it will likely impact on agriculture's ability to feed the world.

Table 4.3 Examples of mortality rates across the world

Somalia	137
Mali	115
DR Congo	98
Burundi	82
Bangladesh	38
USA	7
Japan	3
Singapore	3

Source: UN

If, despite our intentions, all this has made you gloomy, have a look at what the late Hans Rosling has to say about population; "don't panic" is his key message, and the graphics are great.[4]

Notes

1 The population division of the United Nations Department of Economic and Social Affairs has data on human global population. Web link: esa.un.org.
2 Worldometers has up-to-the-minute global population (and other global) data that you can see changing. This website provides data that are relevant to many of the issues raised in this book. Web link: worldometers.info.
3 Mortality rates for under-fives (per 1,000 live births) developed by the UN Inter-agency Group for Child Mortality Estimation (UNICEF, WHO, World Bank, UN DESA Population Division) can be viewed online. Web link: ow.ly/dnEK309eWeW.
4 Hans Rosling was a Swedish statistician with a number of informative (and engaging) videos on YouTube. You'll also find his talks on TED.com.

5

ACIDIFYING THE OCEANS

We begin with some school chemistry. Because water is so good at dissolving other substances, it is hardly ever pure, and absolutely pure water with nothing in it at all would have a pH of 7. The pH scale is a measure of the acidity of a liquid, with acids having a value lower than 7, and alkalis higher than 7.[1]

Here are some common examples with approximate pH values:

Table 5.1 pH values of some common substances (acids)

Common acids	pH
Strong acids	<0.1
Human stomach fluids	<1
Car battery acid	1
Lemon juice	2
Fizzy drinks	3
Tomato juice	4
Rainwater	5.5

Source: Wiley

Table 5.2 pH values of some common substances (alkalis)

Common *alkalis*	pH
Blood	7.5
Seawater	8
Toothpaste	8
Detergents	10
Antacid/acid reflux medicines	10
Ammonia cleaners	11
Bleach	12

Source: Wiley

The pH scale is logarithmic so that a decrease of one unit represents an increase in acidity of 10 times; hence, fizzy drinks are about 10 times more acidic than tomato juice, and lemon juice 10 times less acidic than car battery acid.

Rainwater has a pH below 7 (it's about 5.5 these days) because it contains carbon dioxide that has been absorbed from the air. But if the rain contains other gases such as sulphur dioxide (SO_2) or nitrogen oxides (NO_2), it can be much more acidic. In the 1980s, there was much concern about acid rain, as it was then termed, because of the high concentrations of these gases arising from industrial pollution. This caused considerable corrosion, ecosystem damage and poor human health until a way was found to capture such gases as they went up power station and furnace chimneys.

Seawater, by contrast, has a pH above 7 (it's about 8) because of the solids that are contained within it. Since the Industrial Revolution, the amount of carbon dioxide (CO_2) has increased in the atmosphere, and about 25% of the CO_2 we now release is removed from the air and absorbed by the top layer of the oceans. Initially, this was thought to be a good thing because it took the gas out of the air, but many studies have shown the absorbed CO_2 is changing the chemistry of the seawater and lowering the pH. Although this process is referred to as acidification, seawater is not actually acidic.[2]

The shells of molluscs, the skeleton of corals and the exoskeleton of some algae are all impacted adversely by the increasing amounts of CO_2 being absorbed by the oceans. But why? Carbon is found in the ocean in three dissolved forms: CO_2, bicarbonate ions and carbonate ions. The pH of seawater is dependent on the proportion of these at any time or place.

Ocean water acts as an excellent buffer that limits the effects of adding substances (such as CO_2) that form acids in water, thereby preventing massive pH changes. It's the bicarbonate and carbonate ions that are responsible for the buffering; the principal mechanism is the reaction of carbonate ions with calcium ions in the water to form solid calcium carbonate ($CaCO_3$), which sinks very slowly to the ocean floor, which means that less dissolved calcium and carbonate ions are available in the water. This harms all those organisms that depend on these dissolved ions to generate their own calcium carbonate in order to produce and maintain the quality of their shells.

Despite this buffering, it is estimated that, since the Industrial Revolution, the pH of surface ocean waters has dropped by 0.1 pH units. This may not sound much, but because of the logarithmic scale, in reality, this is a 30% change in acidity, and it is predicted that this process will continue while we continue to generate carbon dioxide in the ways we do. Estimates vary, but some suggest that by the end of this century the surface waters of the ocean could be twice as acidic as they are now.

As is often the case, there are benefits as well as drawbacks to more carbon dioxide in the surface layers of the oceans. For example, algae may benefit as they require CO_2 for photosynthesis, just like plants on land, but such positive effects can also disrupt food chains and ecosystems in unpredictable ways. Species that will definitely not benefit include oysters, clams, sea urchins, plankton and corals. As such, whole food chains may be affected, along with the billions of people who rely on the ocean for food.

Although many organisms that have calcium carbonate shells or skeletons will be adversely affected by a lower pH in seawater, it is the effect on corals that has caught most attention in recent times, as it is clear that a lowering of pH has significantly reduced the ability of reef-building corals to create the skeletons they need. Research suggest that by the end of the century coral reefs may erode faster than they can be rebuilt, which would reduce their viability and affect the million or so species that depend on them.

Ocean acidification is clearly going to be a global problem, although it is currently not possible to predict how it will finally affect the marine food chain and ecosystems. Only one thing is certain: the only way to prevent what is happening is to reduce and then reverse the amount of carbon dioxide entering the atmosphere. This is one of the things the Paris Agreement aims to do.[3]

Notes

1 For an explanation of the chemistry that underpins acids and bases, see the "Acids, Bases and pH" section of Wiley's Essential Biochemistry website. Web link: ow.ly/s8RD308HsHw.

2 The IUCN website has detailed data on ocean acidification. Web link: ow.ly/vyui308HsND.

3 The Paris Agreement is the UN Framework Convention on Climate Change. You can see how many countries have signed online. Web link: unfccc.int/2860.php.

6

ELECTRICITY GENERATION

It's a warm, bright, breezy August afternoon in the south of England with high-level broken cloud. The country is generating around 27.5 gigawatts (GW) of electricity.[1] This is enough to run 27.5 million one-bar electric fires, and results in about 1.2 tonnes of carbon dioxide being released into the atmosphere every second.

The sources and amounts of this electricity are:

Table 6.1 The UK's electricity generation on an August afternoon

Coal	0
Gas turbines	12 GW
Nuclear	8 GW
Wind	3 GW
Hydro	0.7 GW
Interconnectors	3 GW
Other sources	0.8 GW

Source: National Grid

"Other" is mostly solar and biomass. The interconnections are two-way cables linking the UK electricity grid to those in France, Ireland and the Netherlands, with new ones now being built to Belgium, Denmark and Norway. The really surprising thing about these figures is not the proportion of renewables (wind, solar and biomass), which has grown significantly in recent years, but the absence of coal as a fuel.

Compare the August figures with these from a similar day in March of the same year when the generation was just over 42 gigawatts:

Table 6.2 The UK's electricity generation on a March afternoon

Coal	16 GW
Gas turbines	13 GW
Nuclear	7 GW
Wind	2 GW
Hydro	0.7 GW
Interconnectors	3 GW
Other sources	0.5 GW

Source: National Grid

The difference in these figures (between a cold March and a warmer August) is 15.5 GW – almost exactly the amount of power contributed by coal. Historically, coal was the core fuel for the electricity industry (in the 1970s, about two-thirds of electricity came from it), but it has been gradually replaced by gas because gas-generated electricity produces much less CO_2 than coal, and this fits with our commitment to reducing CO_2 emissions. Now coal is something added to the mix when it's needed, and in 2015 its total contribution came to around 20%. The trends are clear – between 2014 and 2015, the change in consumption was: coal (down 25%), and wind and solar (up 33%). This will likely continue as more coal-fired power stations are shut, with gas, wind and solar filling the gap. This is a global trend, and in the US since the start of 2015, five large coal-mining corporations have asked for bankruptcy protection because of debts brought about by this flight from coal. Whether an American president who says he's friendly towards coal makes a lasting difference remains to be seen.

This is only about electricity, of course. On that same August day, 1 tonne of natural gas was imported into the country every second, to be burnt or stored for peak demand, with some of this generating electricity. The gas comes from the North Sea, and from the Middle East in liquid form. There are also gas interconnector pipelines to Belgium, Ireland and the Netherlands.

Such usage figures change with the season and the time of day, and can shift dramatically from hour to hour, or even from minute to minute, when there are sudden surges in demand. Because of this, those running electric and gas grids have to be able to respond quickly, and anticipate changes. Scrutiny of TV schedules remains a key element of this, not because electricity consumption by televisions is particularly high, but because boiling kettles and pumping water after the flushing of toilets are. However, this is less of a problem than it used to be for those who have to keep the electricity supply and demand in balance as the rise of on-demand and iPlayers means that programme watching is more spread out through the week.

A problem that planners have is that the electricity grids were designed and built at a time when there were a small number of large power stations providing most of the supply, with peak demand being met through gas and other systems (such as pump-storage turbines where water in a lake on a mountain is released to flow through turbines near ground level, with the water being pumped back up at night when there is surplus, and therefore very cheap, power). The UK has two such turbine sets that can be kept spinning using air, and then be working at 100% capacity in a few seconds, generating power very quickly. These are still a vital part of the supply mix.

But rather than having a small number of power sources, there are now lots. For example, at the end of 2015, there were 842,000 solar PV installations with a capacity of nearly 9 GW. However, as the Sun does not shine all the time, 9 GW is a maximum that's never achieved. That's not to downplay its significance, though, as it made a contribution to the 25% of electricity that was generated from renewables in 2015, even if the vast majority of this came from wind.

To help cope with fluctuating demand, the grid can ask power stations to be on standby to generate power, and can also tell producers not to produce; each of these strategies has a cost. In 2012–2013, this was about £170 million, which is about 0.2% of consumer bills.[2] The grid can also slightly reduce the voltage at which electricity is supplied, which means that there is a bit less power in the electricity that we get; that is, it goes farther. In extremis, the grid can also selectively cut off consumers, although these are always likely to be industrial users, and it likes to give advance warning of this. Playing fast and loose with families in this way is deemed too politically dangerous, at least in the richer economies.

The buffer between supply and demand has fallen over time, and in the UK now stands at around 1% spare capacity, down from around 5%, where the effect of a large generator going offline had to be catered for. Uncertainties in the supply of electricity are probably inevitable until there is an electricity grid that covers the whole of Europe and North Africa, with the benefits of the growth of solar and wind generation shared and fully used. The National Grid says: "We expect that new technology, from smart meters to innovative forms of electricity storage, could offer new opportunities for large and small consumers to help us balance the system."[3] This means using large battery stores (not yet built), including those in electric cars, as a source of standby power, as well as having the ability to switch off domestic appliances (such as fridges and freezers) for short periods of time. It's not clear how people will react to this, but it could be that it's a better prospect than unplanned power cuts and other uncertainties about supply. Perhaps there will be financial incentives.

As costs (particularly of solar) continue to fall, however, this will probably help boost the growth of renewables. Even so, it will be a while before unsubsidised solar will cost the same as gas. However, if we were to factor in gas's so-called externality costs (that is, the costs of all the pollution caused and CO_2 emitted) into the price paid by the consumer, that might be much sooner. But no government will dare to do that any time soon. As to the future, if the UK is to meet its 2050 target of an 80% cut in greenhouse gas emissions, it will need a continuous source of zero-emissions power, which implies nuclear and/or carbon capture and storage technologies. Expert opinion has it that something like 20 GW of zero-emissions power and 40 GW of wind power would take us somewhere near to the target. Time will tell if this is possible.[4]

Notes

1 At the Energy Watch website, you can see the proportion of fuels currently being used to generate UK electricity. Web link: ow.ly/W7M4308T1WY.

2 Elexon provides a guide to electricity trading arrangements in the UK. Web link: ow.ly/TFf2308T2KL.

3 The National Grid website explains how the electricity system works. Web link: ow.ly/iaXE308T2tA.

4 Energy Trends shows data on electricity statistics. Web link: ow.ly/ juOe308T2gS.

7

GENETIC MODIFICATION

Processes that change the genetic make-up of plants are as old as history – older in fact, as the evolutionary processes that saw plants evolve began long before we humans came along, populated the Earth, and started to think about plants and the past. We tend to refer to such changes as natural because we have no role in them; they just take place because of random changes in plant genetics, leading to new organisms, some of which thrive, some of which don't. One of the most significant of these changes was when bacteria, which had developed the ability to absorb sunlight and use it to bring about useful chemical changes, were first absorbed into plants. You can still find the echo of these bacteria in the chloroplasts within plant cells that start the photosynthesis process by using the energy within light to split water into hydrogen and oxygen. The oxygen is released and the hydrogen combines with carbon dioxide to form simple molecules that the plant converts into sugars, and then into more complex substances.

It's thought the simplest purposeful human plant breeding strategy of identifying plants with a desirable trait (such as a high yield) and selecting them to grow more plants might be almost 10,000 years old. A much more effective approach, however, is to exploit the genetic exchange that lies at the heart of sexual reproduction to breed plants (and also animals).

This uses techniques to optimise desired characteristics such as higher-yielding or more nutritious crops, resistance to viruses, fungi, bacteria and other pests, tolerance of heat, drought or saline conditions, or better adapted to different growing environments. We have been doing this for a long time as well.

This cross-breeding of plants of the same species with different characteristics can produce new varieties with the desired properties. For example, high-yielding varieties of tall-stemmed wheat can be crossed with short-stemmed ones to try to produce high-yielding short-stemmed wheat, and this has been done. The benefit is that less straw is produced, and the crop is less likely to be blown flat in high winds. This sounds straightforward, but the process can be lengthy and is always unpredictable because of the random nature of this genetic exchange where chromosomes are combined to create the diversity. Gregor Mendel's work on the non-random nature of inheritance helped researchers steer a clearer course, and huge strides have been made in all this. It is now a sophisticated international business that is impossible to do justice to here. Such techniques resulted in large yield increases during the twentieth century, leading to the so-called green revolution across the world in the 1960s.

None of this involves any direct, purposeful human interference with specific genes within plants, which is what genetic modification (GM) does. This uses bacteria to either add or remove a specific gene to produce a desired characteristic, producing what are known as transgenic plants – or genetically modified organisms (GMOs). Supporters of such approaches say that doing this can produce plants with desired characteristics faster than classical breeding techniques, as particular genes are targeted, leaving the majority unaltered. They also say that with such genetic modification, you can be certain what genes have been introduced or removed, which classical breeding cannot determine. None of this impresses the opponents of genetic modification, who tend to say that adding genes from different species (sometimes even from animals), which is what transgenic means, is inherently risky. Some say it's just unnatural. It should be said here that it's also not all that precise because knowing where an introduced gene will end up is not an exact science.

The UK Soil Association is one of a number of organisations that is completely against both genetically modified ingredients in human and animal food, and the commercial growth of genetically modified (GM) crops in the UK.[1] Three of its key arguments are:

- The promise that GM crops will revolutionise farming and solve world hunger through higher yields and improved nutritional value have come to nothing.
- The introduction of pesticide-resistant GM crops has led to more, not less, use of herbicides, and resistant superweeds have been created that trap farmers into buying more pesticide.
- Desirable characteristics such as pest resistance and tolerance of drought, flood and salinity can be produced far more quickly and cost-effectively through classical breeding practices.

Those who can see merit in some aspects of genetic modification say that such root-and-branch opposition risks losing benefits that might help feed the world's hungry. One such person is Mark Lynas, who was once a committed anti-GM activist. Lynas began a talk to the 2013 Oxford Farming Conference like this:

> I want to start with some apologies. For the record, here and upfront, I apologise for having spent several years ripping up GM crops. I am also sorry that I helped to start the anti-GM movement back in the mid 1990s, and that I thereby assisted in demonising an important technological option which can be used to benefit the environment. As an environmentalist, and someone who believes that everyone in this world has a right to a healthy and nutritious diet of their choosing, I could not have chosen a more counter-productive path. I now regret it completely. So I guess you'll be wondering – what happened between 1995 and now that made me not only change my mind but come here and admit it? Well, the answer is fairly simple: I discovered science, and in the process I hope I became a better environmentalist.[2]

Such a public *mea culpa* is rare. In Europe at least, both the science and politics of genetic modification remain in flux, but elsewhere, such as in the USA, things look much more settled, with GM foods being grown and sold, although not labelled. GM supporters, and those reasonably neutral in the debate, say that no one has died from eating the stuff.

The technology, of course, continues to develop, with different approaches to modifying genetic composition. Genome editing is one approach that is precise in that it adds or subtracts pieces of DNA in a way that resembles mutation in the wild, which is at the heart of conventional plant breeding. Another approach is genomic selection, where, with enough computer

power, breeders can quickly learn which individual plants are worth growing and breeding from. Crops produced in this way are now competing
with GM products for farmers' attention. They ought perhaps to be viewed
positively by all those opposing conventional genetic modification for the
inherent threats they argue it embodies. But will they? Well, we shall see,
but this will require the sort of courage that Mark Lynas exhibited.

Meanwhile, over the horizon, there is the idea of developing a much
more efficient version of chlorophyll, the molecule that does the work in
chloroplasts. As this may well involve genetic modification, genome editing, genome selection and who knows what, getting acceptance may be
very problematic if minds remain closed to all new technologies.

Notes

1 The Soil Association, along with a number of environmental pressure
groups, has taken a consistent stance again genetic modification. Web link:
soilassociation.org.
2 Mark Lynas's *mea culpa* during his lecture to the Oxford Farming Conference
in January 2013 can be found on his website. You will also find a range of other
articles about the issue of genetic modification. Web link: marklynas.org.

8

CLEAN WATER FOR ALL

The first target of the UN's sixth Sustainable Development Goal (SDG) is, by 2030, to achieve universal and equitable access to safe and affordable drinking water for all.[1]

That doesn't seem much to ask given that water is the basis of life, and safe drinking water and proper sanitation are two of the foundation stones of human health and wellbeing. But the UN says, for at least 1.8 billion people, it really is too much to expect as they still have to use sources of drinking water that are contaminated with sewage. Currently, around 2.4 billion people have no access to functioning toilets or latrines, and more than 80% of human waste goes into rivers or the sea without being treated. It's completely unsurprising, then, that every day, nearly 1,000 children die of preventable diarrhoeal diseases.

In addition to the idea of its being safe, the sixth Sustainable Development Goal uses three other adjectives in relation to drinking water: universal, equitable and affordable. That is, fairly available to everyone at prices they can afford. This may be as hard to achieve as making it safe, because water scarcity affects so many people, with the UN saying that water scarcity affects more than 40% of the world's population, a proportion that is set to rise. For example, over 1.7 billion people are currently living in areas where water use exceeds the rate at which supplies are being replenished.

Despite there being "water, water, everywhere" on Earth, only a small proportion is fresh water, and much of this is locked away and hard to reach.[2] Of the 1.3 billion cubic kilometres of water on Earth, the oceans account for about 97% of it, covering about 71% of the planet's surface. The rest is in glaciers and ice caps, in rivers, lakes and aquifers, in the atmosphere, in the soils and rocks, and, of course, in us and every other living creature. The website of the US Geological Service has more detail on all this than you will probably ever want, as well as some fine images.[3]

Although we cannot drink the water in the oceans, happily nature helps us out through the water cycle, evaporating the stuff into the atmosphere from where it falls as rain, snow, etc. And that water we could drink as it only contains low concentrations of dissolved gases and minerals. So you might think living in a region with plenty of rainfall would be ideal for ensuring plenty of drinkable water. But that's not the case. The water that falls as rain goes onto land, from where it gets into rivers and lakes and into aquifers, and by the time it's extracted for drinking or other uses it can be contaminated with all sorts of impurities, both mineral and biological, some benign, some dangerous. Thus, it's only fit to drink if it gets purified, which is fine if there is a purification plant near where you live. This is not always the case. In many parts of the world, groundwater is becoming badly contaminated, sometimes with salt, as in parts of Australia, and sometimes with poisons, such as arsenic in parts of Bangladesh.

And then there's the problem that rainfall just isn't reliable and just about anywhere can suffer temporary droughts. Some of this unreliability is even predictable, as when it's an El Niño year, we know that some parts of the world will get less rain than usual, and some will get much more. Oxfam has said that about 60 million people across the world faced worsening hunger and poverty in 2015 due to droughts and crop failures when the world experienced the strongest El Niño weather system ever recorded. It seems clear that climate change will likely make everything even less predictable and more unreliable, as a warming world means more water in the atmosphere, making weather systems more energetic, chaotic and violent. Currently, floods and other water-related phenomena give rise to around 70% of all deaths related to natural disasters, a figure that will probably increase with a warming planet. Another possible climate change issue is that, as temperatures rise, glaciers retreat, reducing the flow of water into some of the world's great rivers such as the Ganges, which millions of people rely on.

Given the water-related plight of so many people across the world, it's no surprise that there are now a number of major international charities that focus on water. A typical example is WaterAid, which has teams working in 37 countries. It emphasises partnerships with governments, civil society and people in what it says about its work:

> Ending extreme poverty and building a healthier, more prosperous world for the poorest people cannot be done without safe water, sanitation and hygiene . . . Alongside women's organisations, health, education and marketing partners, we are delivering effective campaigns to eliminate bad hygiene practices, such as defecating in the open.

Even though the UK sits at the eastern edge of a huge ocean, where rain-laden weather systems normally queue up to crash against our shores, there can be a huge difference in rainfall between the wetter west of the country and the drier east, and prolonged periods of dry weather are not unknown. You only have to listen to gardeners or farmers comparing notes to realise what a difference location makes to what is possible and sensible to grow. But that basic geographical difference is no guarantee against local flooding when too much rain falls too quickly, overwhelming both natural and human-created defences. But many, such as George Monbiot, argue that what we have done to the land has acted to weaken natural defences against flooding, by, for example, cutting trees down in the areas where rain falls, reducing the capacity of the land to hold onto water, thereby speeding up its flow into rivers that become overwhelmed. Monbiot's response is to call for a rewilding of the land; this, he says, would bring other benefits, as well as reducing both the occurrence of floods and their intensity.[4]

The uncomfortable reality is that ours has never been a world where having enough water to drink has been a certainty, or that what water there is will be safe to drink. Such unreliability looks set to continue. What will also likely endure across the world is competition for water. Sometimes, this is between peoples and governments, as in the conflict between Israel and the Palestinians; sometimes, it is between domestic and agricultural use, as in California. Often, it is both. That said, there is also good collaboration over water, with examples from India standing out, illustrating that scarcity does not have to mean dire consequences, if the political will is there to find a solution.

Notes

1 The United Nations web pages on Sustainable Development Goal 6 (ensure access to water and sanitation for all) have data and detail on clean water and effective sanitation across the world, and on the practical steps being taken to increase its availability. Web link: ow.ly/c1dH308J6gJ.

2 Coleridge, S.T. (1834). The Rime of the Ancient Mariner. In: H.N. Coleridge (ed.), *The Poetical Works of S.T. Coleridge*. London: W. Pickering.

3 The US Geological Survey has a Water Sciences School with data on where the Earth's water can be found, in its various forms. Web link: water.usgs.gov/edu.

4 Monbiot, G. (2015). *Feral: Rewilding the Land, the Sea, and Human Life*. London: University of Chicago Press.

9

GENDER DISPARITY

In Chapters 29 and 30, we discuss Amartya Sen's idea that human development is a measure of our ability to lead lives that we have reason to value, but such development is unlikely to be attainable for those who are subjected to discrimination.[1] There is copious evidence to suggest that our gender roles, which, unlike our sex, are a product of family, society and culture, give rise to significant disparities. According to the World Economic Forum (WEF), 2016 was "an ambiguous year for global gender parity," as the gender gap widened in 68 countries while being narrowed in 74. For example, although two-thirds of economically developing countries have achieved gender parity in primary school attendance, in many countries girls continue to face barriers to entering both primary and secondary schools.[2]

There are two main global measures of gender equality. One is the WEF's Global Gender Gap Index.[3] This ranks countries in terms of: (i) political empowerment; (ii) health and survival; (iii) economic opportunity and participation; and (iv) educational attainment. Data have been gathered since 2006, and are used to indicate the gaps in outcomes between women and men rather than measuring inputs (for example, the money spent). According to WEF data, no country has completely closed the gender gap, although Iceland is the most equal, with other Nordic countries close behind. The widest gap reported by the WEF is in Yemen. Some countries have made great efforts to reduce such inequalities; Rwanda, for example, had, by 2016, reduced its gender gap by 80%.

The other main measure is the Gender Inequality Index (GII), which is built along the same lines as the Human Development Index; it measures gender inequalities between countries in three aspects of human development:

- reproductive health – maternal mortality rates; adolescent childbirth rates;
- empowerment – proportion of female members of parliament; proportion of females and males aged 25+ with some secondary education; and
- economic status – labour force participation rate of female and male populations over age 15.

GII is expressed as a decimal number between 0 and 1, where the lower the number, the less inequality there is. Of the 155 nations that submitted full data sets in 2014, Slovenia emerged as the most equal nation, with a GII of 0.016. Once again, Yemen was at the bottom, scoring 0.744.[4,5]

Wealth is a less reliable predictor of gender equality than cultural traits. Qatar, one of the world's wealthiest nations, is ranked 116th (GII = 0.524), chiefly because of the absence of any female members of parliament. Female parliamentarians help Saudi Arabia achieve a GII score of 0.286, putting it just one place behind the US, which is ranked 55th (GII = 0.280) because of its relatively high rate of childbirth among adolescent girls. Given that Saudi Arabia is ranked 141st out of 144 on the WEF Index, this GII score might tell us as much about the dangers of using league tables for complex issues, as it does about actual gender disparity.

Statistics illustrating women's reduced income, status or opportunities play out in our societies as a loss of rights and dignity. In extreme cases, such disparities lead to the condoning of domestic violence and practices such as honour killings and female genital mutilation. Such issues can grab our attention, but risk distracting us from commonplace power imbalances that persist around us. These days, we are told that we are free to make our own choices, but, as Cordelia Fine reminds us, choice is not the same as empowerment.[6] Our preferences are formed largely by the society in which we live, and Fine is concerned that the claim of choice is used to justify "a status quo in which politics, wealth, science, technology and artistic achievement lie predominantly in the hands of white men."

This mention of white men reminds us that discrimination can come from different directions. This is called intersectionality, and means that having certain attributes (for example, being poor, female, disabled, etc.) exacerbates other existing inequalities. A famous case was that of black women working at General Motors in the US in 1976 who lost their claim of gender discrimination because black male workers suffered similar disadvantages. Similarly, they failed to prove race discrimination because female white workers were also paid less than their male counterparts. A more contemporary example of intersectionality occurs in sub-Saharan Africa, where the poorest girls are almost nine times less likely to ever set foot in a classroom than the richest boys.

Clearly, development cannot be deemed sustainable if it is achieved at the expense of women's capabilities to realise their own aspirations and participate fully in society. In many traditional societies, women have a triple role – domestic, economic and community. In modern societies, this continues to be the case, yet this contribution often goes unacknowledged when our preferred measure of development is confined to economic wealth. Just as our economies tend to externalise environmental costs (that is, not taking into account the goods and services that the Earth provides), so we fail to account for the unpaid domestic work, such as caring for children, the sick and the elderly, that sustains our communities. This work continues to fall predominantly to women, yet without it, our market economy could not continue to function. This suggests that efforts to build a greener economy may unravel if attention is not paid to the different implications of development for women and men, especially if such development simply adds to women's care burdens.[7]

Rather than being interpreted as being anti-progress, such gender awareness should ensure that development is inclusive. The benefits of such inclusivity can extend to wider environmental concerns, for example in relation to our growing population. Economic growth coupled with better opportunities for women and girls over past decades has been linked to declines in fertility rates. In other words, there is a significant inverse relationship between increased female education levels and a lower birth rate.

The lesson is clear: ensuring that a section of society is not disadvantaged appears to benefit everyone. As one UN organisation puts it, "You can't win if half the team is on the bench."[8,9]

Notes

1 Sen, A. (1999). *Development as Freedom*. Oxford: Oxford University Press.
2 The UN's Education for All report (2015) reports on progress towards gender parity in primary and secondary education. Web link: ow.ly/gLhU308KQ9i.
3 The WEF's 2016 Global Gender Gap Report can be downloaded from its website. Web link: ow.ly/PH4F308KP7k.
4 GII data are available online. Web link: ow.ly/HRne308KPvj.
5 The UK is ranked 39th in the 2014 Global Gender Gap Index; its GII score was 0.177.
6 Fine, C. (2010). *Delusions of Gender: The Real Science Behind Sex Difference*. London: Icon Books.
7 The UNDP's Human Development Reports can be downloaded from its website. Web link: hdr.undp.org/en.
8 This is a quote from Karen Ellemann. Web link: ow.ly/grAd30aJgkC.
9 The UN web pages on Sustainable Development Goal 5 (achieve gender equality and empower all women and girls) have facts and figures on gender inequality across the world, and details of some of the practical steps being taken. Web links: ow.ly/D6RN308JuE7 and unwomen.org.

10

NATURAL RESOURCE DEPLETION

One definition of a natural resource is that it's a material source of wealth, such as timber, fresh water or a mineral deposit, that occurs in a natural state and has economic value. The key phrase here is economic value. Whether something qualifies as a natural resource depends on two things: whether we *can* do something with it, and whether we want to. It's important to note these are more to do with human society than with the thing itself. Thinking about uranium illustrates this. The Earth's crust contains uranium ore in various forms. This is now a valuable resource for use in power stations and in atomic weapons. However, before the discovery of radioactivity, the development of the atomic theory and our ability to use uranium, it wasn't a resource because we couldn't do anything with it. That is, even though we might have known it existed, it was of no use to us, and so not a resource. Were we ever completely to abandon nuclear power and give up atom bombs, uranium would once again cease to be a resource. There are many examples through history of such shifts. Oil is another good example of something that has only fairly recently become a resource. The same applies to the rare earth elements such as neodymium, terbium, yttrium and dysprosium because of their recent novel use in microelectronics.

Because we know that all such ore deposits in the Earth's crust are finite, we know that we might one day have used them all up, and concerns about this (usually accompanied by dire warnings) have cropped up from time to time. For example, in 1931, the US geologist Charles Leith warned that some of the world's most valuable minerals were being rapidly depleted, and in 1952 the US Materials Policy Commission produced a pessimistic report which said that there'd be an end to certain metal ores in a generation. There is a long history of such pessimism, and the classical nineteenth-century economic models of Thomas Malthus and David Ricardo had an emphasis on resource scarcity because they were both convinced that periods of relative abundance could not last. This concern was broadly accepted by natural scientists such as Charles Darwin in his theories of natural selection, but largely rejected by economists. This split was well illustrated by responses to the 1960s Club of Rome report, *The Limits to Growth*, which was widely (though wrongly) interpreted as forecasting imminent shortages of many metals.[1] Once again, it was largely accepted by scientists but not by economists. This is an extract from *The Limits to Growth* (p. 66):

> The Earth's crust contains vast amounts of these raw materials which man has learned to mine and transform into useful things. However vast those amounts may be, they are not infinite . . . Given present resource consumption rates and the projected increase in these rates, the great majority of the currently important non-renewable resources will be extremely costly 100 years from now. The above statement remains true regardless of the most optimistic assumptions about undiscovered reserves, technological advances, substitution or recycling, as long as the demand for resources continues to grow exponentially.

The Limits to Growth contained data on many resources. For example, it said that the amount of known chromium reserves in 1972 was around 770 million tonnes, of which about 2 million tonnes were being mined annually. This suggested that chromium ore might last for 420 years or so. But it said that if use were to grow, say at 3% per year, then the resource would last for fewer than 100 years.

The Limits to Growth authors were well aware that the size of reserves is not just determined by geology. For example, it's clear that if additional accessible reserves are found, then resources will last longer. However, if prices rise, then low-concentration reserves, or those harder to get at, might become viable, providing, that is, there's enough cheap energy available to extract them at a price that's acceptable. In a sense, we have always been running out of things, only to then discover that we are not, and how you do the calculations about resources, and the assumptions you make, are crucial in all this. Going back to chromium, in 2011, the US Geological Survey said that world resources of one major ore of chromium, chromite, exceeded 11 billion tonnes, which was enough (at current rates of use) to meet world demand for many centuries.

However, warnings about the scarcity of raw materials are now incessant themes in the commonplace argument that says humans are running out of material (stuff), fossil fuels and time. See, for example, the warnings about peak oil; that is, that our use of it has outgrown the industry's ability to find more sources. Generally speaking, resource prices rise because of demand or scarcity, or both, and scarcity might be political in nature rather than geological. It is said by some that, for the last 100 years, we have lived through a long period when real prices, adjusted for inflation, have fallen − although there were, of course, many peaks and troughs to do with wars and economic downturns, respectively.

But these sorts of calculation depend on the timescale used and on what commodities are chosen, as the following illustrates. In 1980, Paul Ehrlich (author of *The Population Bomb*) and Julian Simon (an academic economist) disagreed about whether the price of natural resources would rise or fall over the coming years. Simon challenged Ehrlich to choose any raw material and any timescale of more than a year. Ehrlich chose copper, chromium, nickel, tin and tungsten, and the period of 29 September 1980, to 29 September 1990. Simon won, with all five metals falling in price over the period. However, if the wager had included all economically significant commodities, or if the period had run to, say, 2011, Ehrlich would have won.[2]

It seems inevitable that if our economies continue to use and then discard as waste valuable resources, there will be continuing price rises to reflect that usage. Unless, that is we change the economic model to something such as, for example, a circular economy.[3,4,5] Technological change is also a factor, and sometimes there are twists in the resource story. There are some metals, for example copper, that, having been extracted from underground ores, are promptly buried again. The ground is full of copper wire put there by telephone companies when the metal was the best available means for sending phone messages. But now that microwaves and fibre optic technology are replacing the copper and creating faster connections, all that copper becomes a resource again, and an increasingly valuable one, particularly as it doesn't need refining.

Notes

1 Meadows, D.H., Meadows, D.L., Randers, J. & Behrens, W.W. III (1960). *The Limits to Growth*. New York: Universe Books. Web link: ow.ly/68oq3oasTKo.

2 Sabin, P. (2014). *The Bet: Paul Ehrlich, Julian Simon, and Our Gamble Over Earth's Future*. New Haven, CT: Yale University Press.

3 McKinsey (2016). *Sustainability & Resource Productivity, Number 4* offers ideas about whether – and how – the global economy can be nudged onto a less resource-intensive path toward sustainable growth. Web link: ow.ly/7Akc3o8Spdo.

4 Heck, S., Rogers, M. & Carroll, P. (2014). *Resource Revolution: How to Capture the Biggest Business Opportunity in a Century*. New York: New Harvest.

5 Allwood, J.M. & Cullen, J.M. (2012). *Sustainable Materials with Both Eyes Open*. Cambridge: UIT Cambridge.

11

MIGRATION

The current migration of people into Europe from North Africa, the Middle East and further afield because of war and other social turmoil has already been linked to climate change. This is not only because this has been seen as a contributor to the conflicts, but also in the sense that what we are seeing now is likely to be a harbinger of things to come as the world warms further and greater numbers of people will seek more hospitable (in every sense) places to live.

Migration, however, applies not just to people, but to nature more generally, and a 2016 report from the Royal Society for the Protection of Birds (RSPB), *The Nature of Climate Change: Europe's Wildlife at Risk*, explores the issues.[1] This is part of Mike Clarke's introduction, which lays out the issue clearly:

> We are at a point in recent geological history where the rate of human-induced climate change will far outstrip the ability of species to adapt successfully, especially when the resilience of nature has been reduced by habitat loss, non-native species introductions and over-exploitation. The disruption to the web of life is a threat not just to wildlife, but to the lives of people around the world.

The report sets out evidence that wildlife of all kinds will be challenged because of climate change, and it says that protected areas and nature reserves will be vital in helping wildlife cope with a changing climate. This is a view that many wildlife charities agree with. In the UK, for example, higher rainfall will adversely affect birds such as bearded tits, capercaillies and shags, and warmer temperatures in southern Europe will result in habitat loss for Dartford warblers. But Clarke says that it's not all bad news, as new bird species have already begun breeding in the UK; for example, little egrets, black-winged stilts and little bitterns. Clarke also argues that we may need to be much more welcoming of nature's migrants, and therefore less precious about the idea of native species:

> The wildlife we typically accept as being part of our native flora and fauna is moving, and new species are arriving as colonists, partly driven by climate change. The assemblage of species we consider native is therefore in a state of flux. We cannot arrest the changes, so to aid adaptation it will be important to enable species to colonise new areas via provision of sufficient, suitably-protected habitat, in areas that will become more climatically suitable over time.

This is also part of what Fred Pearce argues in his book *The New Wild: Why Invasive Species Will Be Nature's Salvation*.[2] Pearce says that keeping out non-native species looks increasingly flawed as a strategy, and that we should celebrate their dynamism and the novel ecosystems they create. He argues that in an era of climate change and widespread ecological damage, we should be finding ways to help nature regenerate, and that embracing this new wild is our best chance. Clearly, not everyone will agree with this open, liberal approach, and there are certainly other, considered points of view.

In all this, it's necessary perhaps to distinguish between the invasive and the merely non-native; that is, between those migrant species that are here causing trouble and those that are just here. Non-native species are those found outside their natural range because of direct or indirect introduction by humans (unless they just flew in unaided). Where such introduced species persist in unmanaged habitats, they are termed naturalised. It is obvious that many of these naturalised species are not a problem. However, if they spread and out-compete native species, they can threaten ecosystems, habitats or the existence of native species themselves, and give rise to environmental damage and/or economic cost. It is then that they are regarded as invasive. The grey squirrel is a prominent example in the UK.

Invasive species can be plants, animals, fungi or algae that cause disease or pest problems, and the Royal Horticultural Society (RHS) says that, after habitat destruction, invasive non-native species are the most serious threat to global biodiversity.[3] The RHS says that in 2016 in the UK, there were 1,402 non-native plant species, but that only 108 of these (8%) are considered as invasive and a problem. Internationally, the database of such troublesome species is managed by the International Union for Conservation of Nature (IUCN), which now lists 3,163 plants and 820 animals. In 2016, the EU approved a list of Invasive Alien Species of Union Concern (14 plant and 23 animal) that member states must eradicate where possible. Mercifully, this does not include John Wyndham's triffids, that ultimate invasive species.[4]

One problem is that many migrant species can take a long time to become invasive, and many of the plants now considered invasive have been growing in the UK for over 100 years without causing a problem. Where they are a problem, however, they can be expensive to eradicate, and it can take a long time; for example, at least 10 years might be needed to eradicate giant hogweed from an area, and three to four years to completely get rid of Japanese knotweed.

A 2016 *Economist* article, "Invasive Species: Day of the Triffids" (which surprisingly doesn't mention climate), argues for a measured and pragmatic approach to non-native species.[5] It quotes Chris Thomas, a biologist at the University of York, who has calculated that of the UK's 677 most widespread plant species, 68 were introduced by humans before 1500, and another 56 after that date, with not one of these introduced species ranking among the 50 most widespread plants in the country. Even the notorious Himalayan balsam is so rare that it only just makes the list. This is, of course, to take a national view, whereas all politics (whether about plants or people) is local, and Himalayan balsam is very widespread and has to be tackled wherever it is found.[6]

As we noted at the outset, there are some parallels in all this with the current debate about the migration of peoples, although there are clearly important differences as well. For example, some of the language regularly used in relation to plants and animals cannot in all conscience be used about people. But it's possible that a discussion of the migration of plants and animals, and how tolerant we should be of the benefits and problems they bring, might ease a consideration of the much more difficult topic of the movement of people.

Notes

1 RSPB (2015). The Nature of Climate Change: Europe's Wildlife at Risk. Web link: ow.ly/Wauq6.
2 Pearce, F. (2015). *The New Wild: Why Invasive Species Will Be Nature's Salvation*. London: Icon Books.
3 The RHS has good details of invasive non-native species that are present across the UK. Web link: ow.ly/Wauus.
4 Wyndham, J. (1951). *The Day of the Triffids*. London: Michael Joseph.
5 *The Economist* (2015). Invasive species: day of the triffids. 5 December. Web link: ow.ly/Waulx.
6 The Great Britain Non-native Species Secretariat was set up to meet the challenge posed by invasive non-native species in Great Britain. Its website provides tools and information for those working to support the strategy. Web link: ow.ly/dX1y308L206.

12

AIR POLLUTION

Everyone knows that air pollution is never good for you; it's unhealthy, especially for the young, old and infirm, causing serious lung damage and breathing difficulties, as well as exacerbating heart problems. Most outdoor air pollution comes from burning fossil fuels to produce electricity or heat, or from internal combustion engines. The pollution from domestic coal-burning used to lead directly to smog, which resulted in bronchitis, pneumonia and, for some, death; lead compounds released by petrol engines were particularly damaging to young children's mental development; and acid rain not only eroded stonework and acidified fresh water, but the sulphuric acid it contained was a serious throat and lung irritant. These are still problems in many parts of the world, but are less so in developed economies than they used to be. Today, it is the mixture of nitrogen oxides (NOx), carbon monoxide (CO), and small particles from vehicle (especially diesel) exhausts that are serious air pollutants, although exposure to open burning of fuels such as wood and dung can run this a close second.

But how much of a pollutant has to be present for air to be regarded as polluted? It's obvious that if all the vehicles in a city were electric, there would probably be no transport-generated nitrogen oxide or carbon monoxide gases in the air to cause ill health. But suppose there was a steady increase in diesel vehicles. How many would there need to be before we'd say the air was polluted?

That obviously depends on the rate at which they produce NOx, which varies, so another way of thinking about this is to ask what proportion of the molecules in the air can be NOx before we'd say that the air was polluted. One in a thousand? One in a million? One in a billion? The answer cannot be zero, as some NOx arises naturally through the chemistry of lightning strikes. Anyway, logic says that there is a level below which there is no health hazard. For example, one molecule of NOx in, say, the whole of Paris is unlikely to be a problem, even if you were unfortunate enough to breathe it in.

Given this idea of a threshold, it is possible to work out what the relatively safe level of the gas is, and then say that the level must not be exceeded over a particular period. Generally speaking, this is the approach that is taken. What the EU does typifies this, and for NOx, its limit is an exposure at a concentration of 200 µg (micrograms) of the gas in every cubic metre of air for a period of one hour, or 40 µg per cubic metre if you expect to be exposed to the gas for a year, which is more appropriate if, say, you live on a street with heavy traffic. One µg is a millionth of a gram and 40 µg of NOx per cubic metre of air is a concentration of 22 parts per billion (by volume). That is, in every cubic metre of air, there should be the equivalent of no more than 0.02 cubic centimetres of NOx. This doesn't sound a lot, but NOx is a nasty substance that not only causes respiration problems itself, but reacts in bright sunlight with other chemicals such as solvents released by industry or ammonia emissions from agriculture to form secondary pollutants, including ozone and sometimes what's known as photochemical smog, which is even more harmful.

The way that the European Union (EU) tries to enforce this limit is by specifying how much emission is allowed to be generated by vehicles, and Regulation 715/2007/EC specifies emission limits for all important toxic pollutants. The NOx emission limit set for new diesel passenger cars and light vans sold in the EU in 2017 was 80 mg of the gas per kilometre driven. Despite this, and the growth in use of electric and hybrid cars, EU countries find it hard to stay within these limits. This is partly because there are a lot of old vehicles on the road that have much dirtier engines than new ones, but it is also caused by the fact that there are just so many vehicles on the road, despite encouragement to use public transport. None of this is helped by companies that lie to the public and to regulators about how polluting their vehicles actually are.

It is proving hard to keep NOx levels down. The European Environment Agency reported that, in 2014, the air in 10 EU countries was above national emission limits for at least one of the major pollutants. NOx is a particular problem in Germany, and it is likely that its reliance on brown coal (lignite) for electricity generation is a major factor in this. It is, however, a problem everywhere. In 2016, the UK government was taken to court by ClientEarth, an environmental pressure group, which accused it of consistently breaching the NOx limits in 16 cities and regions, despite a 2015 UK Supreme Court ruling that immediate action was needed to remedy the problem. The government lost the case and agreed to implement more realistic approaches to pollution measurement, and, as The Ecologist reports, to take steps to bring pollution levels down to legal levels by the summer of 2017. This is likely to mean an increase in clean air zones in cities, and the Mayor of London introduced a £10 daily toxicity charge in the capital's congestion zone for cars made before 2005.[1]

The existence of particulate matter in the air is an equally serious pollution problem. The EU has a limit of 25 μg per cubic metre (averaged over a year), although the limit suggested by the World Health Organisation (WHO) is much tighter at 10 μg per cubic metre. Urban populations, especially those living next to heavy traffic, are routinely exposed to higher concentrations than the WHO limit. It is a big problem in parts of Italy, Poland, Slovakia, the Czech Republic and the western Balkans.[2]

The seriousness of all this is shown by a 2016 report from the Royal College of Physicians and the Royal College of Paediatrics and Child Health in the UK. They said that outdoor air pollution was a major contributor to around 40,000 early deaths a year. In China, air pollution is reported to contribute to the deaths of 1.6 million people every year. This is unsurprising perhaps, as it is said that being in Beijing on one of their bad air days is equivalent to smoking 40 cigarettes. Following all this, there was a report in 2017 of a Canadian study (published in The Lancet) that suggested there was a link between air pollution from traffic and an increased risk of developing dementia.[3]

The UK Royal Colleges report also said that it was important not to ignore the possibility (and seriousness) of indoor air pollution. The biggest problem there remains tobacco smoke, but there are also contributions from wood-burning stoves, faulty cookers and boilers, the solvents in cleaning products and air fresheners, the glues and other chemicals in carpets and composite furniture, and mould and mildew. Radioactive radon gas in areas with lots of granite rocks is also a problem if there is poor ventilation. In fact, all internal air pollution can be reduced by opening windows and letting in some outside air – providing, of course, that the outside air isn't too polluted.[4]

Notes

1 ClientEarth are activist lawyers who are committed to securing a healthy planet. Details of the ClientEarth focus on climate, health, energy, oceans, forests, business and democracy can be found on their website. Web link: clientearth.org.

2 The EuroStat website has a lot of data on air pollution. Web link: ec.europa. eu. Type "air pollution" into the search box. You will find EU air quality standards on the same website.

3 *The Lancet*'s study of "Living near major roads and the incidence of dementia, Parkinson's disease, and multiple sclerosis" can be found on its website. Web link: thelancet.com. Type "living near major roads" into the search box.

4 Public Health England is an executive agency of the UK Department of Health with a mission to protect and improve the nation's health and to address inequalities. Its website is not user-friendly, but here's a link to research on particulates and health. Web link: ow.ly/711V308JwCp.

13

ELEPHANTS, RHINOS AND DONKEYS

In early 2016, a census of African elephant populations was carried out for the International Union for Conservation of Nature (IUCN), an authority on the status of the natural world and the measures needed to safeguard it.[1] The census contributed to IUCN's African Elephant Status Report, which was launched at the 2016 CITES meeting in Johannesburg (CITES is the Convention on International Trade in Endangered Species).[2] The BBC's headline on this report was "Grim outlook for elephants" and its lead news item was that the elephant population had declined by about 111,000 over the past 10 years, with around 30,000 to 40,000 now being killed every year by poachers working for organised crime networks.

The picture from the census was indeed grim. For example, elephant populations have declined by 22% in Angola since 2005 and by 53% since 2011 in Mozambique. The census also recorded carcass ratios, where a ratio of 10% means one dead elephant was seen for every 10 live ones. These ranged from 0.5% in Uganda to 85% in Zambia, with typical figures being around 20% to 30%. High ratios are taken to be evidence of poaching, and IUCN's conclusion was that overall, Africa's elephant population was experiencing its worst decline in 25 years, mainly due to poaching, although habitat loss remains a significant factor as well.

Overall, it's reckoned that there are still about 415,000 elephants in Africa, although this figure could be an underestimate owing to the difficulty of counting them. Of these, southern Africa has the largest number – about 293,000 (70%). Eastern Africa has around 86,000 (20%) and central Africa about 24,000 (6%). West Africa only has approximately 11,000 (under 3%). IUCN says that eastern Africa is the region most heavily affected by poaching, with a more than 60% decline in Tanzania since 2010. Despite this, numbers have been stable or increasing since 2006 in Uganda, Kenya and Rwanda.

Poaching has not yet had the same impact in southern Africa. Although population declines have been observed in Mozambique, major populations in Namibia, South Africa and Zimbabwe are stable or increasing, and there is evidence of elephant range expansion in Botswana. The elephants in central Africa have been massively affected by poaching since the 1990s, and West Africa's elephant populations are now mostly small, fragmented and isolated. While the international demand for ivory continues, it's thought that poaching will as well.

The same picture can be painted for the rhino populations, for much the same reasons. According to the Save the Rhino charity, 100 years ago there were some 500,000 rhinos in Africa and Asia, but this had fallen to around 70,000 by 1970.[3] Save the Rhino says that by 2016, there were only 29,000 left in the wild. In Asia, both the Sumatran and Javan rhinos are listed as critically endangered, as is the African black rhino, but this population across eastern Africa recovered from around 2,300 in 1993 to between 5,000 and 5,500 in 2016. The southern white rhino has fared rather better. The population is said to be around 20,000, although increased poaching since 2008 is a serious threat to all rhino species.

Donkeys have never been high on any list of endangered species, but a number of African countries have put export bans in place to prevent the export of animals to China, where they have long been a key part of rural life as beasts of burden and transport.[4] It's not that China needs more donkey power these days, but now that there are 50% fewer donkeys in the country because of its rapid industrialisation, there is a shortage of donkey skin. This is valuable when boiled as it produces *ejiao*, a form of gelatin, which is part of popular Chinese tonics and medicines with the power, it seems, to cure coughs, relieve insomnia and revitalise the blood. While all this might seem pretty insignificant when compared to the plight of the rhino and the elephant, such shifts can have marked economic consequences. In Niger in 2016, for example, the price of donkeys rose from \$34 to \$147, which prompted an export ban. The response in more economically liberal and globalised Kenya was to see the opening of a dedicated donkey slaughter-house for the Chinese export market.

What to do about poaching divides both conservationists and those with interests in trade. These divisions were apparent at the 2016 CITES meeting, where over 3,500 people met and where 152 governments voted on 62 proposals to change the CITES protocols. As you might expect of a conference devoted to international trade in endangered species, not everyone there was a conservationist. Alongside such people from international agencies and NGOs were hunters, animal rights activists, scientists and politicians who all wanted to save endangered species from extinction, although not for the same reasons. For example, private sector owners of herds of rhino argued that creating a legal trade in horn would stop the criminal gangs, while conservation groups said that legalising the trade would likely lead to extinction.

A significant part of the business of the conference was focused on elephant ivory and rhino horn. Most elephants are listed under Appendix I of CITES, which prohibits all trade in animal parts. In southern Africa, however, they fall within Appendix II, which allows some regulated trade. Despite this, there is still a complete ban on the sale of ivory, although Namibia, Zimbabwe and South Africa want to allow limited ivory trading. They argue that their elephant populations are large, growing and causing problems. But 12 African countries (supported by Sri Lanka) attempted to have the elephants in these southern African countries moved to CITES Appendix I (from Appendix II) to prevent the possibility of trade. This was narrowly rejected in a vote. Swaziland tried to change the CITES protocols to permit a limited regulated trade: (i) in white rhino horn that has either been collected in the past from natural deaths or recovered from poached rhino; and (ii) to allow horn to be harvested in a non-lethal way from a limited number of white rhino in the future in Swaziland. This was also turned down. Late in 2016, the Chinese government announced that it would outlaw the domestic sale and processing of ivory by the end of 2017. As China has the world's largest ivory market, this might just make a difference, if the ban is both implemented and enforced.

Private rhino owners, some of whom own well over 1,000 animals and have tonnes of horn stockpiled ready for sale (like fingernails, horn grows back), want a legal trade as well. They say that this would deter poachers, and one rhino farmer is quoted in *The Economist*: "I breed and protect rhinos. That's what I do. And I think that's what we need to do to save them." Inevitably, not everyone agrees. Many conservationists think a legal trade in ivory and horn is inherently risky.[5] They fear that the criminal product might turn out to be cheaper than the legal one. Further, effective marketing might lead to increased demand and even more poaching. Those who support what they term the sustainable use of wildlife, by which they mean regulated hunting and trade, say that unless governments allow this, the animals will die out in the wild. Conservationists, meanwhile, say that will happen unless governments listen to them.

Meanwhile, in the middle of all this are the African people who live alongside these animals, many of whom are very poor. *The Economist* quotes Ross Harvey, a researcher at the South African Institute of International Affairs, who says it's important that conservation should provide them with economic benefits. Otherwise, he says, saving endangered animals "is going to be seen as a very middle-class issue."[6]

Notes

1 Details of the 2016 Great Elephant Census can be found on its website. Web link: greatelephantcensus.com.

2 IUCN's African Elephant Status report can be downloaded from its website by typing "African Elephant Status report" into the search line. Web link: iucn.org.

3 The Save the Rhino website has data on the current rhino population. Web link: ow.ly/tCit308Q2PQ.

4 Allison, S. (2016). The great African donkey rush. *Daily Maverick*, 9 September. Web link: ow.ly/V8k9308Q33g.

5 *The Guardian* (2016). Debate: would a legal ivory trade save elephants or speed up the massacre? 1 October. Web link: ow.ly/eBCk30at40H.

6 *The Economist* (2016). To sell or not to sell? 29 September. Web link: ow.ly/AcdX308Q3w4.

14

ENVIRONMENTAL JUSTICE

On 11 February 1994, the President of the United States, Bill Clinton, issued Executive Order 12898, which directed all US federal agencies to develop a strategy that "identifies and addresses disproportionately high and adverse human health or environmental effects of its programs, policies, and activities on minority populations and low-income populations." The executive order required the US Department of Justice to promote the equal enforcement of all existing civil rights, health, and environmental statutes.[1]

Environmental justice is defined by the US Environmental Protection Agency both in terms of fair treatment, but also in relation to the meaningful involvement of people in decision-making – all people, regardless of race or ethnicity, gender or sexuality, origin or income. The Clinton executive order was felt necessary because of mounting evidence that poor, disadvantaged and marginalised people across the US had the odds stacked against them when it came to: (i) ensuring necessary environmental goods, such as clean air and water; and (ii) avoiding a whole lot of environmental bads, such as bad housing, poor food, pollution of all kinds, but particularly nuclear and toxic wastes, hazardous industries, and injurious practices.

The US is far from alone in facing such problems, although not all countries are keen to face up to them. Generally speaking, the more open and democratic a country is, the more likely it is that environmental justice policies will be in place, along with actions to promote them. The European Union has only recently taken environmental justice seriously, and even now ClientEarth, a group of campaigning lawyers, claims that no individual or campaign group has ever been given permission to bring an environmental case against an EU institution to the European Court of Justice.[2] This is despite the right to be able to do this being enshrined in the UN Economic Commission for Europe's Aarhus Convention.[3]

Environmental justice is a broad term and covers many aspects of life. While air, soil and water pollution are significant environmental justice issues, so are poor access to healthy affordable food, inadequate affordable transport links, the ready availability of fuel, and protection from environmental disasters such as earthquakes and tsunami. Famine and drought are ultimate environmental justice issues. It's generally poor people who face such problems, as wealth usually allows people the scope to do something about the issue; often, this is the ability to move house if necessary to a geographical area not so affected. Something else that enables people to resist the imposition of problems, whether by government agencies or business, is the ability to self-organise at community levels and fight back. In this sense, environmental justice is a particular kind of social justice. At the launch of a new United Nations Environmental Justice Council, the UN Environment Programme's Deputy Executive Director stressed this point, saying:

> Social justice must go hand-in-hand with efforts to improve environmental sustainability. With its expertise and global reach, the new council will be a powerful global advocate for law, justice and good governance. It will give renewed impetus to efforts to secure the solid legal foundations on which an inclusive, low-carbon, resource-efficient green economy can be built.[4]

The US's environmental justice movement is dominated by issues of race inequality, but campaigns around the world have different targets. For example, in the UK, the main issues are poverty and the environment, along with health inequalities and social exclusion.

More widely, the Environmental Justice Foundation has sought to make a direct link between the need for environmental security and the defence of basic human rights.[5] The foundation works internationally to protect the environment and such rights, and its campaigns have led to governments addressing forced child labour and human trafficking. Successes also include bans on pesticides, the arrests of illegal (pirate) fishing trawlers in west Africa, and protection for coastal forests in Brazil. They have also campaigned against cotton production in Uzbekistan that relies on child labour, and the mismanagement of water resources for crop irrigation. The foundation has successfully petitioned large retailers to stop selling Uzbek cotton.

The University of Michigan has many case studies of communities across the world dealing with these issues. These include the effects of copper mining in the Philippines, the building of the Three Gorges Dam in China, Maasai land rights in Kenya and Tanzania, Aboriginal rights in Australia's Northern Territory, and the displacement of indigenous people in Malaysia.

At the very highest level of international law, we have the Rome Statute signed by 122 nations (as of 2015). This focuses on "the most serious crimes of concern to the international community as a whole" (also known as Crime Against Peace). Currently, there are four of these:

1   The Crime of Genocide;
2   Crimes Against Humanity;
3   War Crimes; and
4   The Crime of Aggression.

Since the 1970s, the understanding that ecological damage can threaten our peace and security has led to numerous UN committees discussing whether there should be a fifth environmental Crime Against Peace, and in 2010 the environmental lawyer Polly Higgins submitted a proposal to amend the Rome Statute to include an international Crime of Ecocide.[6] Higgins' organisation, Eradicating Ecocide, defines this as:

the extensive damage to, destruction of or loss of ecosystem(s) of a given territory, whether by human agency or by other causes, to such an extent that peaceful enjoyment by the inhabitants of that territory has been or will be severely diminished.[7]

The idea behind creating such an offence at the very highest level is that when someone or some entity commits this kind of crime, the International Criminal Court has powers to intervene.

Notes

1 The US Department of Justice is a key part of the government of the United States. Web link: justice.gov/ej.
2 ClientEarth are activist lawyers who are committed to securing a healthy planet. Details of their focus on climate, health, energy, oceans, forests, business and democracy can be found on their website: clientearth.org.
3 The Aarhus Convention establishes a number of rights for the public (both individuals and groups) with regard to the environment. Web link: ec.europa. eu. Type "Aahrus" into the search box.
4 UNEP Press Release. Web link: ow.ly/Ev3030g2Vxc.
5 The Environmental Justice Foundation is a UK-based charity that works internationally to protect the environment and defend human rights. Web link: ejfoundation.org.
6 Higgins, P. TEDx Exeter: Ecocide, the 5th Crime Against Peace. Web link: ow.ly/J631309cBKN.
7 The Eradicating Ecocide website. Web link: eradicatingecocide.com.

15

ECONOMIC GROWTH

Is economic growth a good thing? As we live in societies where it is a major political goal, the answer would seem to be yes. A straightforward definition of economic growth is an increase in the size of the economy; that is, in the production and consumption of goods and services. This means, generally speaking, that a country can only have economic growth by increasing production and/or consumption (per head of population), or increasing the population – or both. It also comes through innovation, which leads to different forms of goods, services and consumption. Just think how microelectronics has changed the economy and our lives.

Governments get anxious when gross domestic product (GDP) figures are about to be published. These are a measure of the production and consumption of goods and services (that is, the size of the economy), and media analysts pour over the data looking for trends and insights. Generally speaking, positive GDP growth is cheered (and the more positive, the better), and negative GDP shifts are frowned upon. Consistent growth ought to mean more and/or better-paid jobs. But two periods of negative GDP change mean a recession where the economy contracts and jobs are lost or become worse paid. It's because we think that more jobs, and more well-paid ones, are better than fewer or less well-paid jobs that we tend to cheer economic growth along.

There are, of course, a few problems with the idea of GDP as it's not necessarily a measure of what's socially useful. One way to boost GDP (at least in the short term) would be to arrange for every car in the country to have a reasonably gentle crash. There would be no personal injury, and the amount of money subsequently spent on metal bashing would give the economy a bit of a boost. Then next month, we could turn to window smashing or spray-painting buildings to give the economy another nudge. All these silly suggestions make the point that the nature of an economy, and the sort of jobs it contains, is just as (or more) important as its size or how it changes. A serious problem with GDP is that it mainly measures economic transactions but ignores social and environmental costs and inequality measures. Robert Kennedy once said that a country's GDP measures everything except that which makes life worthwhile.[1,2]

There are also problems with the idea of economic growth itself as it tends to mean more and more stuff being produced, all of which also takes additional material (and energy) to generate it. All this places more pressure on increasingly scarce resources, and usually means more waste along the way. This is not only true of agricultural, extractive and manufacturing industries, but the service and information sectors as well. In one sense, more stuff would be fine if it could be produced without needing more new resource material – and the waste and pollution that goes with its extraction, manufacture and use. This is a key idea of the circular economy. But, it's not just a growing shortage of raw materials that's a problem; so are polluted air and water, the loss of biodiversity and peace and quiet, and climate change. All these are the results and indicators of what Herman Daly says should really be termed uneconomic growth because of the bigger, global picture.[3] This problem is all the worse because these negative effects of economic activity (they're called externalities) are usually not costed into the prices paid by consumers. The taxpayer is left paying the bill, even though they might not have had much, if any, benefit.

Some argue that it's not GDP growth that's the real problem here, but the fact that greater economic activity leads to more greenhouse gas emissions. If only, they say, these could be uncoupled, then growth might not be so much of a problem. The measure of this connection is called carbon intensity. This sees changes in greenhouse gas emissions not in absolute amounts, but in terms of GDP.[4,5] According to the 2014 Low Carbon Economy Index report from PricewaterhouseCoopers (PWC), the fall in greenhouse gas emissions in the US (per dollar of GDP) was 1.9%.[6] In Germany, it was a fall of 7.1%, and in the UK 10.9% (the highest in the world). The global average figure was a fall of 2.7%, which comes about because the global GDP rose by 3.3% in 2014, while total carbon emissions rose by (only) 0.5%. As a result of this, PWC suggests that GDP and emissions have at last become uncoupled, which is some rare good news. However, PWC then used data from the UN's Intergovernmental Committee on Climate Change (IPCC) to show that this average figure needs to be 6.3% to keep the world temperature rise below 2°C. Further, they say that to avoid 2°C of warming, the global economy now needs to decarbonise at 6.2% a year every year from now until 2100. This seems unlikely to happen.

Are there any alternatives to this way of thinking about the economy? Is it possible to get away from the idea of only having just two alternatives: increasing or decreasing?[7] Some say that a steady-state economy is what we need. This is one where we no longer strive for growth at any price, but try to have economies that are sustainable in every sense. In other words: a stabilised economy that is neither growing nor shrinking, but fluctuating around a sustainable level. Just how this is to be managed, and by whom, and what the implications are for jobs, are only some of the questions that are asked of this idea. The Centre for the Advancement of the Steady State Economy (CASSE) calls for the development of steady-state economies with stabilised population and consumption to begin in the wealthiest nations, and to do so without people being coerced. As yet, a start has not been made on this.[8]

Notes

1 *The Economist* (2016). The trouble with GDP. 30 April. Web link: ow.ly/ WB613oasQQQ.

2 Constanza, C. et al. (2014). Development: time to leave GDP behind. *Nature*. Web link: ow.ly/I94w308L3Ag.

3 Zencey, E. (2015). Economics ignores thermodynamics. *The Daly News*. Web link: ow.ly/HsrJ308L3Ib.

4 Randers, J. (2012). *2052: A Global Forecast for the Next Forty Years. Report to the Club of Rome.* New York: Global Foresight Books.

5 Heinberg, R. (2010). *Peak Everything: Waking Up to the Century of Declines.* Gabriola Island, BC: New Society Publishers.

6 PriceWaterhouseCooper (2014). *Two Degrees of Separation: Ambition and Reality. A Report on Its Low Carbon Economy Index (LCEI).* Web link: ow.ly/ lKVs308L3tC.

7 E.F. Schumacher came to believe that our use of technology was not in balance with human need. He wrote *Small Is Beautiful: A Study of Economics* as if people mattered, and founded the Intermediate Technology Development Group. A short film about his life can be viewed online. Web link: ow.ly/9T4y309cwX0.

8 The Centre for the Advancement of the Steady State Economy. Web link: steadystate.org.

16

WATER POLLUTION

In the 1960s, Michael Flanders and Donald Swann, two English musical satirists, wrote a song about "the many interesting things you may come across in the countryside, if you're not careful."[1] Titled "The Bedstead Men," its theme was that, in the country, if you come across a pool or pond, you will always find some rubbish in it, most often "a big, brass broken bedstead." It was a comment on both the state of Britain's waterways at that time, and on Britain itself.

Brass bedsteads are long out of fashion. For a while, their place in Britain's waterways was taken by supermarket trolleys. Though even this is now a rare sight as technology can largely prevent trolley theft. However, just because water has no physical objects floating or sunk in it doesn't mean that it's not polluted, as so many substances dissolve in or mix with water that it can be hard to tell how clean it is. Water pollutants include: detergents, food waste, insecticides and herbicides, fuels and lubricants, industrial solvents, cosmetics, pharmaceuticals, industrial waste, heavy metals, acids, nitrates, and phosphates.

Fertiliser run-off into rivers, causing eutrophication, is a particular problem. This is where a body of water becomes enriched in dissolved nutrients (such as phosphates). This overstimulates the growth of aquatic plants, usually resulting in the depletion of dissolved oxygen in the water. This, in severe cases, reduces water quality, affecting fish and other animal populations. And then there are the many organisms that can be injurious to health and live in water, even in developed economies with all their regulation and safeguards. These include campylobacter, cryptosporidium, salmonella, giardia, E. coli, norovirus and parasitic worms. Across the world, the threats are even greater from problems such as malaria, typhoid, dysentery, cholera, tapeworms and bilharzia. Guinea worm has disappeared from this list recently thanks to a successful international effort to control it – based on filtering water.

Even the best drinking water is never pure in the sense that it's only H_2O. As your local water company will readily tell you, there are inevitably minerals in it, although there are regulatory limits to resources such as lead, aluminium, copper, iron, arsenic, fluoride and nitrate that water might contain. The more sophisticated the water treatment, the more control there will be on such minerals, many of whose limits are only a few parts per billion.

But such problems pale almost into insignificance if you haven't got piped drinking water or effective sanitation, as is still the case for many across the world. Although over 1.9 billion people have gained access to piped drinking water since 1990 and over 2 billion now have improved sanitation, making this more comprehensive still is a focus for the 2015 Sustainable Development Goals, where Goal 6 is to ensure availability and sustainable management of water and sanitation for all.[2] The following figures show the scale of the challenge:

- Some 2.4 billion people lack basic sanitation such as toilets or latrines.
- At least 1.8 billion people use a source of drinking water that is contaminated by sewage.
- Almost 1,000 children die every day because of preventable water- and sanitation-related diseases.
- More than 80% of human waste water is discharged into rivers or seas without any treatment.[3]

UNICEF, the UN agency focusing on children, has a programme it calls Water, Sanitation and Hygiene (WASH) that focuses on children's ability to drink safe water, and the journey they must take to collect it.

For sanitation, WASH works to ensure access and use of basic toilets and ways to separate human waste from contact with people, and to end the still widespread practice of open defecation. It also works to nurture good hygiene practices, especially basic hand-washing with soap. UNICEF has highlighted that the opportunity cost from a lack of access to water disproportionately falls on women and girls, who collectively spend as much as 200 million hours every day – the equivalent of more than 22,800 years – just collecting water.

Here are some of the Sustainable Development Goal 6 targets for 2030:

- Universal and equitable access to both safe and affordable drinking water, and to adequate and equitable sanitation and hygiene.
- An end to open defecation, with particular attention paid to the needs of women and girls and those in vulnerable situations.
- Improved water quality through reduced pollution, minimising the release of hazardous chemicals and materials, halving the proportion of untreated waste water, and substantially increasing recycling and safe reuse.

Much of this is amenable to the application of proven and cost-effective standard technologies, and a further goal is to expand international cooperation to support good-quality water and sanitation across the world. Examples include water harvesting, desalination, water efficiency, waste water treatment, recycling and reuse technologies. A further goal is to support and strengthen the participation of local communities in improving water and sanitation management. The last point is important as political will and cultural acceptance can be important in the adoption of even simple technologies.

Given these challenges, it will be a good day when it's possible for everyone to wake up and only worry about the possibility of bedsteads, trollies and other detritus in the world's waterways, and not whether there is enough water nearby that's fit to drink.

Notes

1 Flanders and Swann's satire, "British Bedstead Men," is available on YouTube. Web link: ow.ly/xAQI3o8SBBy.

2 The United Nations web pages on Sustainable Development Goal 6 (ensure access to water and sanitation for all) have data and details on clean water and effective sanitation across the world, and on the practical steps being taken to increase its availability. Web link: ow.ly/c1dH308J6g.

3 The World Health Organisation (WHO) has wide-ranging data on water and its availability across the world. Web link: who.int.

17

EATING MEAT

For a long time, and across many (but not all) cultures, eating meat has been seen as something to aspire to because it implies wealth and status, and because meat is a concentrated form of easy-to-absorb protein, vitamins, minerals and energy that are essential to a good diet. The idea of meat and two veg (which usually includes potatoes, pasta, etc.) as the core of a balanced diet is of long standing, and up to a few years ago, meetings of those interested in the environment had to pre-specify vegetarian food to cater for those oddball characters who didn't eat meat. Now, to attend a meeting almost anywhere, is to find vegetarian food as standard. This is because more people *are* vegetarian (or vegan) these days, and more (perhaps many more) are just eating less meat, or eating it less often. This may be because of a recognition of the high environmental costs that meat-eating involves, because of an ethical choice about animal welfare, or because of a personal concern to eat a healthier diet.

That said, there is still a great deal of meat eaten across the world, and, as people in economically developing countries get richer, their instinct can be to eat more meat because they can afford to. Research by Chatham House, a UK think tank, using UN data, says that by 2050, compared to

2014 levels, global consumption of meat is expected to rise by 76%, and dairy products by 65%.[1] In 2015, the three biggest meat-consuming areas were China, the EU and the US, with about half the meat consumed being pork. Unsurprisingly, there are good nutritional arguments in favour of having meat in some form in your diet; one is the need to get enough Vitamin B12, which is essential for the normal function of nerve cells, DNA production and blood. This is found naturally in meats, poultry, seafood, eggs and dairy products, especially blue cheese, but if you don't eat any of these, supplements are necessary. Chatham House says that the global meat industry generates more greenhouse gas emissions (14.5% of the total) than all forms of transport put together. Cattle account for about two-thirds of the emissions from livestock, and this, compared to the production of soya protein, generates around 150 times as much greenhouse gas, by volume. Even rearing low-emissions meat such as pork and chicken results in up to 25 times more greenhouse gas than soya does.

Many now argue that reducing our consumption of meat is essential if we are to avoid devastating climate change. The fifth assessment report of the UN Intergovernmental Panel on Climate Change (IPCC) argued that such dietary change can substantially lower carbon emissions because the livestock industry is the largest global source of two potent greenhouse gases: methane (CH_4) and nitrous oxide (N_2O).[2] Livestock production also results in forest loss and increased carbon dioxide (CO_2) emissions as forests are cut down to provide pasture, or are degraded through animal grazing. Meanwhile, rising demand for animal feed means the expansion of crops into cleared forests. Despite these warnings about emissions, governments continue to subsidise meat production. Across rich countries, this cost taxpayers $53 billion in 2013, with cattle subsidies in the EU exceeding $731 million, about $190 for every cow. In China, pork subsidies now exceed $22 billion, about $47 per pig.[3,4]

For many people, there is an efficiency argument here as well. These days, with modern techniques, 10 times as many people can be fed on an area of land given over to grain, pulses, etc. than if that land is used to rear animals. Oddly, this is not an argument for having no animals at all, as there is value in having a cow or pig on your land for the direct fertilising and soil conditioning that it provides. In the Middle Ages, having an animal and enough area for crop rotation normally meant your land could be kept fertile and productive. Much the same conditions still apply in many parts of the world today, and there is a strong argument that a truly sustainable agriculture will involve animals, and that these will eventually enter the food chain.

The ethical argument, which is of long standing, is an objection to killing sentient animals for food when we do not need to do this to have a good diet. These are matters of principle and values. In this sense, over half a million animals are needlessly killed in UK slaughterhouses every week. An allied objection is the poor standards of animal welfare that can be found in today's intensive, industrialised meat industry, and in animal transportation and slaughter. Further, it is increasingly hard to know very much about what is really in our food, where it has come from, or what animals have been fed on. Sometimes food labelling is misleading or simply untruthful, as various horsemeat scandals confirm.

There are also persistent concerns about hormonal growth stimulants and antibiotics in the animal food chain, and about bacteria and toxins remaining in meat carcasses. Although millions of Americans seem happy to eat chlorine-washed chickens, they're not everyone's cup of tea. Many such problems arise from the drive to produce meat cheaply so that more people will be able (and want) to buy it. There are also recent concerns that a diet rich in red meat slightly increases our risk of cancer, and the core message from many health authorities in developed economies now is: reduce your meat-eating to help you live a healthier and longer life. In the UK, this is accompanied by the five-a-day campaign to encourage the consumption of more vegetables and fruit, although the choice of the number five turns out to have been rather arbitrary.

Although subsidies and taxes affect our food habits, we still do have a choice about what we eat, and how much we care about animals, ourselves and the planet. Michael Pollan argues that although meat-eating may be full of moral and ethical ambiguities, it is possible to consume meat with a clear conscience if we know that the animal has lived, fed and died well.[5] Happily, this would be good for our health as well, particularly if eating good-quality meat meant eating less of it, as it probably would. If so, the pressure on the planet might diminish a little, and all-round misery be reduced.

Notes

1 Chatham House has published a number of studies of meat-eating, particularly in relation to climate change. Web link: chathamhouse.org. Type "meat eating" into the search box.

2 The fifth assessment report of the UN's IPPC can be read online. Web link: ow.ly/8pbK3o8QrRn.

3 *The Meat Atlas*, published by the Heinrich Böll Foundation Germany and Friends of the Earth Europe, can be seen online. Web link: ow.ly/8wpu3o8 Qrmp.

4 The UN FAOStat helps to achieve food security for all, making sure people have regular access to enough high-quality food to lead active, healthy lives. Web link: faostat3.fao.org/home/E.

5 Pollan, M. (2007). *The Omnivore's Dilemma: A Natural History of Four Meals.* Harmondsworth: Penguin.

18

WEATHER AND CLIMATE

Although we are used to our weather changing – sometimes more than once a day – a changing climate is another matter. Although climates do change, the rate at which they do this, historically, is reasonably slow. For example, there is evidence of a warmer period across the North Atlantic (roughly from 800 to 1200 CE), and of a cooler period from about 1300 to 1850 CE. The first enabled Norse people to settle and farm southern Greenland; the second made Frost Fairs possible on the River Thames.

When we first began to think seriously about the Earth's getting warmer (global warming), and the effect that this might have on the planet and on our lives, a determined effort was made to change the language that was used to talk about such things. The problem with the phrase global warming, the argument went, was that people might think it was a good idea, and so not be worried by it. Who in northern Europe, for example, wouldn't like it to be just a little bit warmer in the winter, and have a bit more sunshine in summer? And so, the phrase climate change was introduced to help persuade us to take the idea more seriously. Wishful thinking apart, the problem with the phrase global warming is easy to see; although the Earth might get warmer, this will not necessarily result in the weather we'd like: much colder winters and very wet summers are as much a possibility as the opposite. Indeed, it seems likely that a warmer world would mean even more unpredictable weather than we are used to.

But the use of the term climate change was itself controversial, and a new group soon emerged of climate change sceptics or deniers. Such people either said that the climate wasn't changing, accusing scientists of fiddling the data, or asked, given that climate changed naturally, what was the evidence that human activity was the cause. Surprisingly perhaps, these groups got on well together. In response to some of these arguments, the language was adjusted again, and the idea of rapid climate change was introduced.

But what is climate? The climate you have depends on where you are on the Earth, and might be summed up as the long-term weather pattern in a region, particularly in terms of sunshine, temperature, wind (direction and speed) and rainfall or snow, and how these vary on a seasonal and yearly basis. But sunlight, temperature, wind and rain are determined by your latitude and altitude, and by your nearness to oceans and mountains, and it is these that are the key factors in determining the sort of climate we experience. They explain why the west of the British Isles tends to experience warm(ish) south-westerly winds that bring a lot of cloud and rain, whereas the south of France, for example, is usually much sunnier and warmer. While some of these factors do change over very long periods, for example because of the movement of the continents, there must be more to a changing climate than this.

There are some predictable shifts that contribute to regular changes to climate; for example, changes to the Earth's tilt, and to its orbit around the Sun. These affect the amount of sunlight we get, and where it falls. Others are less predictable; for example, how much heat energy is held within the atmosphere and in the ocean, how the dominant ocean currents change, and how this alters the transportation of heat around the Earth and its exchange between land and water. As an example of the difficulties, we do not yet fully understand the natural El Niño and La Niña effects, even though they cause weather to change across the planet, often dramatically so. However, in all this, it seems clear that the greenhouse gases in the atmosphere are important in determining how warm it is and what happens to climate.[1,2]

The Earth is warmer than it ought to be, given its distance from the Sun, because of these greenhouse gases, but how warm it actually is has varied quite widely over time. Historically, there have been several very cold periods – ice ages – when temperatures dropped so low that sea ice covered the polar regions, and glaciers lay over much of the land in northern latitudes and Antarctica. Sea levels fell because so much water was held as ice on the land. When warming reoccurred, the glaciers melted (we picture them as retreating) and sea levels rose, leaving altered coastlines and changed landscapes. Ice ages occur when either the amount of the Sun's radiation falling onto the Earth, or absorbed by it (or both), drops. The temperature of the atmosphere is affected by how it holds heat energy, and it also depends on how much of the heat gets transferred to the oceans in a natural (but complex) mechanism. The last ice age, when temperatures fell by about 12°C, and the sea level by 130m, ended about 15,000 years ago, but the changes to the Earth are not over. For example, the British Isles are slowly tilting (north-west rising, south-east sinking) as the land very slowly adjusts to the loss of the huge weight of ice.

And so, is our climate changing? Well, the UN's Intergovernmental Panel on Climate Change (IPCC) says it is. It notes that since the 1950s, there have been unprecedented changes to the climate: the atmosphere and ocean have warmed, the amounts of snow and ice have diminished, and sea level has risen. Its 2014 report shows that each of the last three decades has been successively warmer at the Earth's surface than any preceding decade since 1850, and that the period from 1983 to 2012 was likely the warmest 30-year period of the last 1,400 years in the northern hemisphere. The IPCC says it is extremely likely that more than half of the observed increase in global average surface temperature from 1951 to 2010 was caused by increases in human-produced greenhouse gases. It also says that it's probable that these have also affected the global water cycle, contributed to the retreat of glaciers since the 1960s, and to the increased surface melting of the Greenland ice sheet since 1993. It also says that these gases have also very likely contributed to Arctic sea ice loss since 1979, making a substantial contribution to the increases in sea level rise observed since the 1970s.

Climate change, the IPCC says, is already affecting drinking water quantity and quality, and many terrestrial, freshwater and marine species have shifted their geographic ranges, seasonal activities, migration patterns, abundances and species interactions in response. There have been more negative impacts on crop yields than positive ones, and some impacts of ocean acidification on marine organisms. The IPCC says that changes in many extreme weather and climate events have been observed since about 1950, and that some of these have been linked to human influences and activity. Further, it is very likely, it says, that the number of cold days and nights has decreased (and warm ones increased) on the global scale. It also says that it is likely that the frequency of heat waves has increased in large parts of Europe, Asia and Australia. It adds that it's likely that storm surges have increased since 1970, being mainly a result of rising mean sea level. The IPCC is careful with its language. When it says "extremely likely," it means greater than a 95% probability, whereas "very likely" means more than a 90% probability, and "likely" only means more than a 66% probability.[3]

You will likely have your own views on all this, and what you make of what the IPCC says may depend on how you view its methodology, or how much of a political process (as opposed to a scientific one) you think it is. It is also likely to depend on the mental frame you use to consider such matters, as well as on the media outlets you prefer to follow. This is because the media, and not just those strongly against the idea of climate change, will be using a very careful framing to get their ideas across.

Notes

1 NASA's website has a good introduction to climate variability. Web link: ow.ly/KAFc308QtFl.

2 The NOAA website has a good explanation of El Niño and La Niña. Web link: ow.ly/VIw2308QtWR.

3 The UN's IPPC website has a lot of data on climate. Web link: ipcc.ch.

19

OZONE AND CFCs

Ozone is a Jekyll and Hyde molecule. Although it is dangerous at ground level, at high altitudes it protects us from the battering the Earth gets from radiation in space. The ozone absorbs a lot of this ionising radiation, preventing it from reaching the ground and damaging both us and the rest of the natural world.

Ozone is a form of oxygen where there are three atoms in the molecule rather than the usual two. It's written O_3 and is a pale blue gas with a smell rather like bleach. Its concentration in the atmosphere is very low (about 0.00006%). By comparison, oxygen is 21%. Ozone is mostly found above 10 km in the stratosphere where it is formed by the action of ultraviolet (UV) radiation on oxygen. An ozone layer was first suggested in the 1930s, but it took 20 years for research to show that this ozone might be destroyed if it came into contact with the wrong sort of chemical. It took another 20 years before it became clear that something as common as chlorine could do this.

It's the extra atom in ozone that makes it very reactive as it easily breaks up into a normal oxygen molecule (O_2) and a single atom (O). This makes it useful in industrial processes where oxidation is needed, such as in bleaching, in removing odours, and in killing microorganisms in air and water. Here, it's a good alternative to chlorine, and is even advertised as "nature's cleaner" as oxygen is its only by-product.

This oxidising ability, however, causes damage to animal respiratory systems and plant tissues, if concentrations rise to around 0.001%. This makes ozone a pollutant at ground level. It's unfortunate, therefore, that it is produced when UV radiation (in bright sunlight) acts on a mixture of oxygen and exhaust gases from motor vehicles to produce photochemical smog. Because ozone can also be produced by electrical discharges in the air, it can be present wherever these occur, including in factories and offices. Although ozone is a strong greenhouse gas, its contribution to global warming is low as there is so little of it, compared to CO_2, and because, as it is very reactive, it does not hang around for long. It used to be thought that the health benefits of exposure to bracing sea air was because of ozone. But if you think that can smell something like ozone, it's because of a chemical generated by phytoplankton in the sea. That said, some people still think that very small quantities of ozone are good for you, and ozone machines are available for use at home.

All was well for the balance between oxygen and ozone in the atmosphere until 1928 when Thomas Midgley greatly improved the process of making a class of chemicals called chlorofluorocarbons (CFCs), which the Belgian scientist Frédéric Swarts had first created in the 1890s. These were molecules containing carbon, hydrogen, chlorine and fluorine atoms and were very useful as they were good solvents for oils, did not burn, and were replacements for the toxic gases ammonia and sulphur dioxide as refrigerants. Industrial uses for CFCs developed over the next decade, and, after the Second World War, the demand for them grew rapidly, especially once their use as aerosol propellants became clear. For a time, they were thought to be miracle chemicals because of the range of uses and the absence of drawbacks.

In the early 1970s, James Lovelock, of Gaia fame, detected CFCs in the atmosphere for the first time. Alarm bells started ringing in 1974 when Frank Rowland and Mario Molina predicted that CFCs would gradually decompose, release chlorine into the stratosphere and damage the ozone layer. In the same year, the chairman of the DuPont chemical company, which made CFCs, noisily rubbished this idea. However, in March 1977, the United Nations Environmental Programme (UNEP) held the first international meeting to address the issue of ozone depletion, and in March

1978 the US banned all non-essential use of CFCs in aerosols.[1] Despite this, there was a continued international expansion of their use, and in 1982 low ozone concentrations were measured over Antarctica. In 1984, the British Antarctic Survey discovered what became known as an ozone hole, but which was, in reality, a thinning of the layer. This discovery led to decisive international action, which we explore in our final chapter on the Montreal Protocol.

Note

1 UNEP's Ozone Secretariat has an online handbook of the Montreal Protocol. Web link: ow.ly/3kMv308QKVA.

20

BIOCENTRISM

As we end the first part of the book, this seems a good place to ask: How biocentric do you think you are in what you do? That is, when you think about humans, and how we live on the Earth along with the rest of nature, to what extent do you think of nature's interests as opposed to just the human ones? Perhaps you never consciously think about them at all; perhaps they are always in your mind. It's likely that most of us will float somewhere on the spectrum between these extremes.

But none of this is straightforward, and how we feel and act might be highly contingent on the circumstances and the context. Being human, some say, means that we are bound to think of ourselves first and foremost, and there is much evidence to support that view. But some of that will be self-interest as opposed to what's in the interests of humanity more widely. Charity, after all, is said to begin at home, and perhaps our charitable giving says something about how we see this balance of interest between humans and the rest of nature? For example, how do you split your donations between people-focused charities and nature-focused ones, between, say, the likes of the RSPCA and charities such as Oxfam? And how is it split between national charities and international ones? Further, if you support animal charities, are these one that concern wild animals, or domesticated ones? All this might say something about how biocentric you are. But it won't be the whole story.

Being anthropocentric (human-centred) is usually seen as the opposite of being biocentric (non-human nature-centred), except that these are not complete opposites. For example, even the most biocentric human is unlikely to forego a course of antibiotics if death threatens, or think that the Ebola virus should be left unchecked to do its worst. And even the most human-centred person is likely to have some feeling for the rest of nature, even if it's only to appreciate a view or the scent of a rose.

A modern definition of biocentrism is that it's an ethical perspective which says that all life has equal moral standing. Historically, ethics were anthropocentric, with only living humans being worthy of such moral consideration. Over time, however, this was extended to future generations of humans, and then to certain so-called higher animals, and thence to all animals and then to all plants, and following that to ecosystems, populations and species; thus, to all of life itself. The argument for this is that by being alive, something is deserving of moral consideration. But a butterfly is one thing, and the malaria parasite quite another. In some ways, none of this is new. It is a key aspect of Buddhism to avoid harming anything living; St Francis of Assisi might be said to have preached a biocentric theology, and some communities leading non-material lifestyles clearly try to live by such ethics.

But how do you live a practical biocentric life? Probably it's only possible with great difficulty and many compromises. Biocentrism puts four duties on anyone trying to hold to it:

1   Do no harm to other living beings.
2   Don't interfere with other beings living their lives.
3   Don't use other organisms as merely means to human ends.
4   Make restitution if a living being has been harmed by human activity.

Viewed strictly here, only particular forms of vegan living would seem even remotely possible – for example, the eating of fruits, seeds, nuts, etc., where the plant is not killed – although how you gather these without interfering with the plant in question is a key consideration as any form of agriculture would seem disallowed. Eggs might similarly be impossible to gather. It seems to describe a particular form of hunter-gatherer living where all you do is gather what nature makes available to you. And that's before we even think of making clothes and the likelihood of lice.

But this is perhaps to miss the point. If the four duties become guides rather than rules, then a lot of people already live by thinking about them. For example, these might readily be modified:

1   Do no harm to other living beings *unless they threaten you*.
2   Interfere *as little as possible* with other beings living their lives.
3   Don't use other organisms as merely means to *trivial* human ends.
4   *Within reason*, make restitution if a living being has been harmed by human activity.

But this is a slippery slope that can soon lead you to a new set of guides:

1   Do not be cruel to other living beings.
2   If you interfere with other beings, do so in a limited way and as ethically as possible.
3   If you use other organisms for human benefit, do so ethically.
4   Where practicable, make reasonable restitution if a living being has been harmed by human activity.

And so on, until you are in danger of flipping the whole thing over to become so non-biocentric that you become as anthropocentric as everyone else.

A strong criticism of biocentricity is that it places too much emphasis on the individual, and in doing so misses the bigger picture. This argument says that it's the population and the species that matter, not individual members of it, although that's not an argument most of us would ever use about humans outside of warfare. This is the sort of argument that justifies removing invasive or non-native migrant species from a particular environment. It leads to a more holistic view that ecosystems and populations are more important than individuals, and that killing individuals can be a useful strategy. Thinking like this justifies the culling of individual animals in order to benefit the herd. Of course, it's humans that think like this, not, as far as we know, the animals.[1,2,3]

Notes

1 DesJardins, J.R. (2013). *Environmental Ethics: An Introduction to Environmental Philosophy*. Independence, KY: Wadsworth Cengage Learning.

2 Johnson, L.E. (2011). *A Life-Centered Approach to Bioethics: Biocentric Ethics*. Cambridge: Cambridge University Press.

3 Taylor, P. (1986). *Respect for Nature: A Theory of Environmental Ethics*. Princeton, NJ: Princeton University Press.

PART II

CONCEPTS

21

THE ANTHROPOCENE

Most people, unless they accept a literal reading of ancient scriptures, think the Earth is around 4,570 million years old, and geologists divide its history into a series of eras, periods and epochs. These are large chunks of time, typically many millions of years. The following table shows a number of the better-known ones, at least one of which has slipped out of geology and into common usage in the south of England's Jurassic Coast, and the *Jurassic Park* film franchise.

Table 21.1 Examples of eras, periods and epochs in Earth history

Title	Duration (*years*)
Holocene	11,700 to the present
Pleistocene	2.6 million to 11,700
Eocene	56 to 34 million
Cretaceous	146 to 66 million
Jurassic	200 to 146 million
Permian	299 to 251 million
Carboniferous	416 to 359 million
Precambrian	4,570 to 542 million

Source: Wikipedia

These timescales are how we humans think about the past, based on scientific evidence to be found in rocks, and in the fossils they contain. Distinctions between these have to be based on clear changes that can be seen in the rocks. For example, the Carboniferous epoch, when the coal-bearing rocks were formed, is obviously quite different from what came before.[1,2]

In 2000, Eugene Stoermer and Paul Crutzen proposed that we should consider the current epoch, the Holocene, to be at an end, and recognise that a new one, the Anthropocene, has begun.[3] The Holocene started at the end of the most recent ice age when the glaciers began to retreat. Before this, during the Pleistocene, there were at least 20 cycles of glacial advance and retreat, and, as much of the world's water was locked up in the ice sheets, deserts expanded and it was cold and dry. Humans evolved in Africa during the Pleistocene.

In the Holocene, as the ice retreated, the soil re-emerged and forests spread, only to shrink again with human demand for timber and land. The case for the Anthropocene – the human epoch – is that we modern humans (Homo sapiens) are now not only living on the Earth, as we have done for 200,000 years or so, but we are fundamentally changing the nature of the planet and how the biosphere functions. Stoermer and Crutzen said that the evidence for this is all around us. And this is something that we have tried to illustrate in the book.

In the summer of 2016, the International Geological Congress was told by Colin Waters, the secretary of its Anthropocene Working Group, that the group thought that the time had come to accept that the shift to the Anthropocene had taken place, and that this should be formally recognised.[4] But when this transition happened remains a vexed question, and there are various possible timings; for example, the proportion of the Earth's land being used for agriculture passing 50%, the movement of species between Europe and the New World, the start of the Industrial Revolution, the making of the first plastics, or the world's first electronic computer. That said, the Working Group seems to favour our Promethean ability to build and use nuclear weapons as the decisive marker, as the testing of these weapons in the atmosphere left a telltale residue of plutonium and other radioactive elements across the planet. By this measure, the Anthropocene is now at least 50 years old, and the shift is being widely acknowledged. For example, it is the core idea within WWF's 2016 Living Planet Report, and WWF says that the word captures both the fact that human activity now affects Earth's life support system and the responsibility we now must shoulder because of this.[5]

But the science is not yet settled. There will be much geological bureaucracy to struggle through if the Anthropocene is finally to be verified by the International Union of Geological Sciences at some point in the future. And it is by no means certain that it will be recognised, as whether we have shifted to the Anthropocene is as much a political question as a geological one. There are those, for example, who see the Anthropocene claim as one more act of hubris committed by our anthropocentric culture, something that sets humans apart, yet again, from our environment.

This move from one epoch to another, Holocene to Anthropocene, is not an arcane point or a trivial matter. To start with, it will be the first time such a shift has happened. More importantly, however, it is not a shift to be welcomed, unless the challenge it offers to politicians everywhere to address its implications in both policy and practice are taken up. In that sense, it fits well alongside the Paris Agreement on climate change and the Sustainable Development Goals.

A question for the very far future is: What will geologists find when they dig around, apart from that telltale plutonium? Will it be fossilised plastic water bottles in the rocks? Will it be a loss of shells in seabed sediments because of ocean acidification? And will this have been replaced by well-preserved plastic waste that currently floats around the oceans? Or will it be layers of all those microbeads that are used in cosmetics or the fibres that wash out of clothes made from synthetic materials? Or maybe it will be new sorts of rocks and minerals made from all the metals that escape recycling and end up in landfill, perhaps with soot and cinders from power stations, and bits of concrete mixed in? Or, more mundanely, will it be, as some suggest, the bones of the humble chicken, now the world's most common bird and favourite food, with some 125 million tonnes eaten every year?[6] That's an awful lot of bones, and they have to go somewhere.

Notes

1 Wikipedia has a straightforward introduction to geological timescales. Web link: ow.ly/mrZj308QLlo.
2 The BBC website has a basic introduction to the history of life on Earth. Web link: ow.ly/Npn7308QLIU.

3 A historical development of the concept of the Anthropocene is provided by the Stockholm Resilience Institute. Web link: stockholmresilience.org.

4 The Anthropocene Working Group papers can be found online. Web link: ow.ly/OpdT308QMoA.

5 WWF (2016). *Living Planet Report*. Web link: ow.ly/THlX308LVLN.

6 For further details of meat consumption, see *The Meat Atlas*, published by the Heinrich Böll Foundation Germany and Friends of the Earth Europe. Web link: ow.ly/8wpu308Qrmp.

22

NATURE

In *My Fair Lady*, when Henry Higgins says of Eliza Doolittle, "Her smiles, her frowns, her ups, her downs; are second nature to me now,"[1] he is using nature to mean something that is innate or intrinsic to a person. Higgins has become so used to having Doolittle around that she now seems part of who he is: someone whose presence is so long experienced and deeply ingrained to be part not just of his life, but of him. We still use the word nature in this sense. For example, we say, "it's not in her nature to do that." Here, we mean that it's not the sort of thing that she ever would, or ever could, do because of the sort of person she is. This is a very old use of the word.

Our more common use of the word nature, to mean the living world of the biosphere, came much later. In a narrow sense, it stands for everything biological: from the smallest components of a cell, to whole organisms, ecosystems, habitats, and beyond these to the whole Earth and the idea of Gaia. As scientific thinking developed in the sixteenth and seventeenth centuries, this meaning of nature came to dominate our use of the word. But nature must relate to more than biology because all living things depend on non-biological systems. For example, we all need water, oxygen and minerals to live; thus, the physical world of ocean, atmosphere and land, and all the materials, resources and forces they contain, are also part of nature – of what we also call the natural world. In this way, a granite outcrop is as natural as a tropical forest, an ocean wave as natural as a dolphin, and a howling gale as natural as a butterfly.

We find it hard to think clearly about nature, and what is natural, because we have a tendency to want to see natural things as beneficial to us; that's why we speak of Mother Nature, and why the words nature and natural crop up so often in advertising. If it's natural, companies want us to believe it must be good; just think of how often you see the phrase "natural goodness" on packaging. But the smallpox virus is natural, as is the plague bacillus, and the toxins produced by clostridium botulinum. And so are volcanoes, earthquakes, tsunami, hurricanes, typhoons, ice ages, and the regular assault by meteorites from space.

Clearly, nature takes an even-handed approach to human wellbeing, and we understand this through two seemingly opposite ideas. The first embodies intense competition; Tennyson wrote of nature as "red in tooth and claw," which captures predator–prey realities in the natural world.[2] The second embodies cooperation, and this captures the way that organisms collaborate with each other for mutual benefit. We might like one idea better than the other, and prefer it, for example, as a way of visualising the ideal human society, but both are real and the issue is how to maintain a dynamic balance between them. Nature does this itself through its own regulatory mechanisms, unless humans come along and badly disrupt them.

To add to our confusions, we often use the word nature, and the phrase the environment, as though they mean the same thing, although we also sometimes differentiate between them carefully; we have environmental education in schools, for example, rather than nature education. But if you are old enough, you may remember nature studies and nature walks, which illustrates that fashion and fad may have a role here. But this idea of the environment is quite different to how we understand nature. In differentiating between these ideas, Stephen Gough says it's helpful to see the phrase the environment as a "container-term for meanings that people attach to those phenomena that are external to themselves," and nature as "the entities and forces that exist independent of such meanings – including ourselves."[3]

In this way, we can conceive of a natural world (nature) that is outside of us and quite independent of how (or whether) we think about it. But the environment is not external to us in this way. It is internal because how we understand it, and what it means to us, depends on how we think about it. As Gough notes, if there were no people on Earth, water would keep falling from clouds because of gravity, but this would not be part of the environment because there are no humans to make it so – and there would be no words for Earth, water, clouds, gravity, etc., because there would be no words at all.

We humans *are* here, of course. We are embedded in nature and we have a sophisticated understanding of it. It follows that how we organise ourselves is similarly embedded; thus, our societies and economies are part of nature. It provides for us in every way, and we are utterly dependent on it for our food and water, our air, and all the resource materials that we use to create the civilisation (and chaos) around us. However, we do not always act as though this were true. Jared Diamond has chronicled how a number of social communities over time have struggled (and often failed) to survive when the way in which they thought about their world (nature) was at odds with the reality, or with how this was changing.[4] Such societies, such as the Norse people who lived in Greenland from around 900 to 1450 CE, were severely handicapped because they did not have much understanding of certain aspects of nature – particularly of atmosphere, ocean and climate – and so could not understand what was happening as the climate changed and the way of life that they valued first got tough, and then became impossible.

We have fewer such excuses for the mess we are making of things today. Whether you think of the pollution of air, land and water, species loss and habitat destruction, the loss of stratospheric ozone, the acidification of the oceans, and the threat of catastrophic climate change, the pressure that human development is placing on nature's ability to keep supporting us in the way to which we have become accustomed is greater than ever. And yet our thinking about nature, and about our dependency on it, is beset with problems because we persist in seeing these as someone else's responsibility rather than our own.

Notes

1 *My Fair Lady* (1964). Directed by George Cukor. Quote taken from the song "I've Grown Accustomed to Her Face".
2 "Nature, red in tooth and claw" comes from Alfred Lord Tennyson's "In Memoriam" (LVI). The verse is: "Who trusted God was love indeed / And love Creation's final law / Tho' Nature, red in tooth and claw / With ravine, shriek'd against his creed." Web link: ow.ly/YMM1308SS9z.
3 Gough, S.R. (2015). *Education, Nature, and Society*. London: Routledge.
4 Diamond, J. (2005). *Collapse: How Societies Choose to Fail or Succeed*. Harmondsworth: Penguin.

23

GAIA

In ancient Greek mythology, Gaia (or Gaea) was the goddess of the Earth who created herself out of chaos at the dawn of creation. She was viewed as the mother of everything, including all the other Greek gods, with all mortal creatures being born of her flesh. The ancient Greeks saw the Earth as a flat disc surrounded by a river with the solid dome of heaven above. The disc rested on, and was inseparable from, Gaia's breast.

This idea of Mother Earth or Earth Mother is very old and is widespread across the world's cultures. Mother Earth embodies nature, motherhood, fertility, bounty and creation (and, it has to be said, sometimes also destruction). The Mother Earth idea still appeals to people now, and is especially beloved by many environmentalists who, while they might otherwise be quite irreligious, can find in the Earth a spiritual quality possibly missing from their lives. Their argument goes: if the Earth is literally the source of all goodness and nourishment (both spiritually and materially), should she not be respected, nurtured and generally looked after? Well, indeed she should, although this is not the only way of thinking about such things, as James Lovelock illustrated when he developed his Gaia hypothesis with Lynn Margulis in the late 1960s. They argued that the Earth is a single organism that actively maintains the conditions necessary for its survival.

In a 1975 article for *New Scientist*, Lovelock and Sidney Epton posed two questions: (i) Do the Earth's living matter, air, oceans and land surface form part of a giant system that could be seen as a single organism?[1] (ii) Could human activities reduce such a system's options so that it is no longer able to exert sufficient control to stay viable?

The first of these lies at the heart of the Gaia hypothesis; the second is the existential threat we now pose to ourselves and to life itself. In their *New Scientist* article, Lovelock and Epton contrasted two propositions:

1   Life exists only because material conditions on Earth happen to be just right for its existence.
2   Life defines the material conditions needed for its survival and makes sure that they stay there.

Proposition 1, Lovelock and Epton say, is the conventional view that temperature, oxygen levels, humidity, ocean acidity and salinity, etc. fall within limits that mean that they are right for life to exist. Proposition 2 is the Gaia view, which implies that "living matter is not passive in the face of threats to its existence. It has found means . . . of forcing conditions to stay within the permissible range."

Inherent in Proposition 2 is the idea that the components of the biosphere (air, oceans, ice, land) are in a kind of dynamic balance that maintains the sort of homeostatic condition that the human body manages with respect to temperature, blood pH, glucose levels, salinity, etc. For example, our temperature is controlled very close to 37°C even though it might be −20°C or +45°C outside. For Gaia, this means controlling within a narrow range atmospheric oxygen and carbon dioxide levels, ocean and soil pH, surface temperature, etc.

In many ways, the idea of Gaia, with its proposition that what most of us regard as inanimate could in fact be thought of as somehow alive, remains as provocative as it was in the 1970s.[2,3,4,5] Since that time, the hypothesis has given rise to many new areas of research about the Earth's physical, chemical, geological and biological processes, and these continue today. However, although it's useful to think of the Earth in systems terms, that doesn't mean it has to be the sort of living system that Lovelock first outlined, and many would now agree with Toby Tyrrell, a professor of Earth system science, when he says that the Gaia hypothesis is not an accurate picture of how our world works.[6]

That said, the idea and image of Gaia remains in the popular imagination with, for example, suggestions that we might view it in physiological terms: the Earth's oceans and rivers being its blood, the atmosphere its lungs, the land its bones, and living organisms its senses. Although this sort of imagery seems too literal a view of what is a sophisticated idea, Lovelock and Epton did end their *New Scientist* article with this:

> Now for one more speculation. We are sure than man needs Gaia, but could Gaia do without man? In man, Gaia has the equivalent of a central nervous system and an awareness of herself and the rest of the universe. Through man, she has a rudimentary capacity, capable of development, to anticipate and guard against threats to her existence. For example, man can command just enough capacity to ward off a collision with a planetoid [asteroid] the size of Icarus. Can it then be that in the course of man's evolution within Gaia he has been acquiring the knowledge and skills necessary to ensure her survival?

Well, perhaps. But this leads us to wonder whether, if we are Gaia's central nervous system, we might be in the grip of a serious meningitis-like viral infection. It was William Golding, the author of Lord of the Flies, who suggested the name Gaia, and we wonder whether he had in mind that she had a reputation as something of a troublemaker among the gods.[7] We certainly think that it's by no means clear that a living Gaia would be eternally tolerant of a species – we humans – that constantly defied her both by its brute carelessness and its hubris. More reasons, if we needed them, to mend our ways.

Notes

1 *New Scientist* (1975). *The Quest for Gaia*, 6 February. Web link: ow.ly/ 6s2Z30804lQ.

2 Lovelock, J. (1979). *Gaia: A New Look at Life on Earth*. Oxford: Oxford University Press.

3 Lovelock, J. (1991). *Healing Gaia: Practical Medicine for the Planet*. Danvers, MA: Harmony Books.

4 Lovelock, J. (2006). *The Revenge of Gaia: Why the Earth Is Fighting Back – and How We Can Still Save Humanity*. Santa Barbara, CA: Allen Lane.

5 Lovelock, J. (2009). *The Vanishing Face of Gaia: A Final Warning – Enjoy It While You Can*. Santa Barbara, CA: Allen Lane.

6 Tyrrell, T. (2013). *On Gaia: A Critical Investigation of the Relationship Between Life and Earth*. Princeton, NJ: Princeton University Press.

7 Golding, W. (1954). *Lord of the Flies*. London: Faber & Faber.

24

BIODIVERSITY

The term biodiversity refers to all life on Earth at every level, from genes through to microbes, fungi, insects, right up to whole ecosystems or communities of living things. The term also covers the genetic differences within each species – for example, between varieties of crops and breeds of livestock – as well as the relationships and processes (such as evolution) that sustain life. Made up from the two words biological diversity, biodiversity became a widely used term after the Earth Summit in Rio de Janeiro in 1992, so much so that you might wonder how we ever managed without it. In the past, we used terms such as natural heritage or species richness, but these weren't so all-encompassing or arresting.

The Earth's ecosystems include wetlands, forests, oceans, rivers, urban areas, farmland and deserts. All these involve numerous relationships, not only among all the different species, but also between the species and the non-living parts of the environment such as water, air and soil. In a presentation to the US Senate, the sociobiologist E.O. Wilson reminded us that:

> within a square foot (of soil) . . . may live . . . billions of microorganisms, and large numbers of species intricately connected with one another and together maintaining that ecosystem in balance. But above all, what is maintained is the biodiversity itself, comprising large numbers of species, each of them hundreds of thousands to millions of years in the making, and in total representing evolution that is exquisitely adapted to that environment.[1]

As well as stressing such interrelationships, Wilson made two other important points: first, that the diversity of species is the result of millions (possibly billions) of years of evolution; and second, that it is the combination of life forms and their relationships with each other and their environment that make the Earth habitable for all life on the planet – and that includes we humans. This is why biodiversity is so important, and you don't have to accept all the Gaia hypothesis to appreciate this.

Plants, animals and ecosystems provide us with numerous goods and services that help to sustain us. They provide food, drink, shelter and medicines – from aspirin to penicillin. They clean our air and purify our water. They may include pests, but they also provide pest control and pollinators for our crops. And it's not just the benefits of individual species, but the complexity of ecosystems, that matters. For this reason, areas of very high diversity such as coral reefs or tropical forests are known as biodiversity hotspots, and are often the focus of international conservation efforts. Such complex systems can deal with new situations because they have a large number of subsystems and components that can help the overall system adapt to change. This is called system resilience. Without it, if there is no way to absorb changes, then the system cannot respond to new realities, and this can lead to catastrophic consequences. Less complex systems such as monoculture in farming or forestry are less resilient, and so they are more susceptible to environmental changes brought about by pests, diseases or climate change. Supporters of organic farming say that it's valuable because it discourages farming techniques that rely on pesticides and herbicides that reduce biodiversity. They argue that modern farming systems with chemical inputs might make farms more productive in the short term, but the real fear is that ultimately such systems may be less resilient.

The Earth is always changing and species extinctions are part of the natural order of things. However, the rate of change has accelerated as humans have come to dominate the Earth over the last few hundred years, and especially so in recent decades. Ecosystems are being altered rapidly, and the planet is facing an unprecedented rate of extinctions. This loss of biodiversity can be through direct causes, such as deforestation and hunting, or indirect ones as a result of widespread pollution or climate change. Coral reefs, for example, are declining because of the oceans becoming more acidic, which harms the organisms that live on – and build – the reefs. This can be a disaster for local fish stocks and numerous other species. That said, efforts are being made to monitor biodiversity loss and protect ecosystems to help ensure the survival of species. The Red List of Threatened Species, maintained by IUCN is the world's most comprehensive inventory of the global conservation status of plant and animal species.[2] Using the best available scientific data, the Red List highlights those plants, fungi and animals that are at risk of global extinction (listed as critically endangered, endangered or vulnerable), and this information can guide conservation efforts locally and globally.

We also have a legally binding, global agreement, the Convention on Biological Diversity, which was adopted at the Earth Summit in Rio. This agreement, ratified by all UN member states (except the US) has three main goals:

- the conservation of biological diversity;
- the sustainable use of its components; and
- the fair and equitable sharing of the benefits from the use of genetic resources.

Since the Convention was adopted, two additional protocols linked to the Convention have been agreed. One is the Cartagena Protocol on Biosafety, which entered into force in 2003 and aims to ensure the safe handling, transport and use of living modified organisms (LMOs) that result from modern biotechnology.[3] This is important because LMOs have the potential to impact negatively on biodiversity, and ultimately human health.

The other is the Nagoya Protocol, which is concerned with access to, and the sharing of, the benefits of genetic resources.[4] This came into force in 2014, and the hope is that by providing greater legal certainty for providers and users of these resources, they will use biodiversity sustainably so that it can continue to contribute to human wellbeing indefinitely.

Notes

1 Wilson's presentation is set out on the Save America's Forests website. Web link: saveamericasforests.org/wilson/intro.htm.

2 IUCN, the International Union for Conservation of Nature, is composed of both government and civil society organisations. It provides public, private and non-governmental organisations with the knowledge and tools that enable human progress, economic development and nature conservation to take place together. Web link: iucn.org.

3 The Cartagena Protocol on Biosafety aims to ensure the safe handling, transport and use of living modified organisms (LMOs) resulting from modern biotechnology that may have adverse effects on biological diversity and human health. Web link: cbd.int.

4 The Nagoya Protocol on Access to Genetic Resources is an international agreement that aims at sharing the benefits arising from the utilisation of genetic resources in a fair and equitable way. Web link: cbd.int.

25

CONSERVATION

The word conservation, and terms like it, are very widely used. For example, in science, we find laws such as the conservation of momentum and the conservation of energy, and there is the phrase energy conversation, which involves measures such as double glazing and cavity wall insulation. In art, textiles and heritage, conservators work on artefacts to prevent their further deterioration or loss; at home, we routinely talk about conserving our energy for tasks that are to come; and conserve is another name for jam. At their heart, these all have the idea of the absence or prevention of loss and maintaining existence and usefulness, although this does not mean that nothing is altered. For example, conserving an artefact usually means that the alterations that are made to it make it more likely to endure. The key idea here is that to conserve something is to maintain its condition in order to protect it against further loss.

Conservation is distinct from conservatism. This is a social and political philosophy that tends to want to retain traditional social and cultural institutions; some seek to preserve things as they are now, emphasising stability and continuity; others oppose such modernist error, and seek a return to the way they think things once were, in the golden age before it all went to the dogs. Conservation also needs to be distinguished from preservation, which means keeping objects as they are now; food can be preserved in aspic, and animal organs in formaldehyde.

In the context of the natural world, the word conservation is widely used and is found associated with ideas such as: nature conservation, habitat conservation, ecosystem conservation, wildlife conservation, water conservation, soil conservation, etc., and organisations exist to promote each of these ideas. The modern nature conservation movement seeks to maintain the quality of the biosphere for the future, so that it can continue to provide support for all life forms – and not just humans. This means having an emphasis on what needs to be conserved: for example, natural capital, biodiversity, species, habitats and ecosystems.

Conservation can be focused at a very local level, in districts, and at national, international and global levels. The UK Wildlife Trusts, for example, are county-based with their own nature reserves. These, although sometimes small, may contain species or habitats that are rare or endangered (or both). The Trusts encourage local people to work with professionals to monitor these reserves and carry out conservation tasks to help nature along. This is important, as most reserves are not at all natural or pristine, in that they have been created, sometimes over many hundreds of years, by human activity. That is, they are now as much cultural as natural.

For example, the English Lake District fells look like they do in part because sheep have been eating bracken and rough grass for the best part of 1,000 years, and the Wiltshire Downs have their distinctive rolling contours because grazing animals keep down hawthorn, rowan and ash. Left to their own devices, such landscapes (and nature reserves) would become places where what we value (the endangered species, for example, or a landscape) has vanished.

Such conservation management tasks include clearing scrub (for example, brambles), removing trees and shrubs that have seeded themselves (for example, blackberry and hawthorn), to taking out invasive plants (such as Japanese knotweed, Himalayan balsam and giant hogweed) and animals (such as mink and signal crayfish). Conservation management also includes using animals to graze grasslands at particular times of the year to encourage certain plants to grow. Such conservation activity can be controversial, especially where it clashes with farming practice or human development of roads, airports, housing and businesses.

Although there are lots of national conservation organisations, many are international. WWF, with its panda logo, credit cards and policy of encouraging the adoption of animals, is one of the most well known. Although best known for its work on the conservation of charasmatic megafauna (the panda, rhino, etc.), it actively works to help people in poor communities lead more fulfilled lives. It does this in part to help prevent pressure that such communities might place on the natural environment, but also because it understands (as do all serious conservation organisations) that environmental problems are caused by what human societies do, and as such, any conservation strategy has to involve people.

WWF played a major role in setting up the international agreement on conservation known as CITES – the Convention on International Trade in Endangered Species of Wild Fauna and Flora.[1] This came into force in 1975 and now extends to 181 countries. Its aim is to ensure that any international trade in wild animals and plants does not threaten the species survival. Such trade (in both live specimens and products) is worth billions of dollars a year and involves hundreds of millions of plant and animal specimens. Today, CITES provides protection to over about 30,000 plant species and 5,500 animal species. Part of the international conservation effort has been the setting up of protected areas, sometimes covering huge areas of land, but these have not been wholly successful, and often exhibit all the problems that zoos have.

Many UK species are protected under the 1981 Wildlife and Countryside Act, and those considered to be in the greatest danger are protected under the 2010 Conservation of Habitats and Species regulations. The very rare large blue butterfly is one example, and it's an offence to disturb or capture them, to take or destroy their eggs or caterpillars, or damage their breeding sites. It's also an offence to have them (dead or alive) in your possession with a view to their sale. It's also a criminal act to release non-native species into the wild anywhere in the UK.

Sites of special scientific interest (SSSIs) conserve and protect the best of our wildlife, geological and physiographical heritage for the benefit of present and future generations. There are over 4,000 SSSIs in England, covering around 8% of the country. SSSIs provide a range of ecosystem services, including:

- cultural services to people and the economy, including tourism, education, sense of place and recreation, as well as clear conservation benefits;
- regulating services, including water purification, regulation of climate, air quality, water and natural hazards, by protecting and enhancing natural processes; and
- provisioning services that they produce, including goods such as food, timber, genetic resources and fresh water.

Natural England is responsible for ensuring SSSIs are managed appropriately and assessing and monitoring their condition. About 80% of SSSIs (by area) are internationally important for wildlife as they are home to the most rare and threatened habitats and species in Europe. These sites are designated as European special areas of conservation, and form part of the network of protected areas known as Natura 2000.

Conservation is local – literally in your own back garden – as well as national and international, and the key to effective policymaking is to find a way of ensuring that laws and practice are mutually supportive. After all, as with climate change, wildlife crime knows no boundaries.

Note

1 CITES is the Convention on International Trade in Endangered Species of Wild Fauna and Flora – an international agreement between governments. Its aim is to ensure that international trade in specimens of wild animals and plants does not threaten their survival. Web link: cites.org.

26

CHARISMATIC MEGAFAUNA

Whenever prominent environmental organisations want more of our cash, they have a habit of confronting us with the plight of what are described as charismatic megafauna – a term coined in 1985 to describe attractive, large, wild animals that could be useful for marketing purposes. These are usually endangered in some sense, and may even be almost extinct in the wild, having to rely on zoos and suchlike for their continuing existence. Of course, such animals are not actually charismatic. They don't really overwhelm us with their personalities, but we undoubtedly find them attractive. Messages to us about their plight tend to be emotional, focusing on our sentimental side, or on our sense of guilt. Such animals are familiar: tigers, elephants, giant pandas, snow leopards, black rhinos, Asian elephants, polar bears, whales, orangutans and mountain gorillas.

However, it's rare for really ugly animals, or ones with bad reputations, to get this coveted status. Crocodiles, for example, Komodo dragons, hyenas and big snakes are rarely included. And woe betide you if you're very small; then you've no chance.

Such animals experience a range of problems in the wild, many of which are as a result of development. The problems include loss of habitat, pollution, poaching, overhunting and competition from other species. There is an official sliding scale of endangerment; here it is with some examples:[1]

Table 26.1 The sliding scale of species endangerment

Extinct	Bali tiger	Barbary lion	western black rhinoceros
Extinct in the wild	scimitar oryx	Père David's deer	black softshell turtle
Critically endangered	Sumatran tiger	red wolf	African wild ass
Endangered	blue whale	bonobo	Ethiopian wolf
Vulnerable	African lion	cheetah	Burmese python
Near threatened	jaguar	American bison	striped hyena

Source: Wikipedia

Such charismatic animals are sometimes known as flagship species. The idea is that because they only thrive in areas of rich biodiversity, helping them can also help many less well-known, less-glamorous but equally endangered species. As such, the argument goes, helping big animals can improve the health of the ecosystem as a whole. It is not clear, however, that this is necessarily the case, as the evidence that such megafauna are indicators of broader biodiversity is mixed, and relying on flagship animals might mean that other species in danger receive less attention. Thus, saving elephants and rhinos will not necessarily help species that occupy significant ecological positions in food chains.

Organisations employ a range of megafauna-based strategies. Sometimes you just get direct appeals for cash. A message might be: if you give us £10 per month, then we can help save the rhino. Sometimes, the word help is omitted. This is also a key component of the strategy aimed to get people to leave legacies to wildlife charities. Sometimes, organisations have adoption schemes whereby giving that same £10 per month gives you a notional link to an animal. For example, the World Wildlife Fund (WWF) offers a range of species to connect with. Through this, in addition to cuddly toys, magazines, bookmarks and stickers, you get information on what the charity is doing to help your particular species. Of course, it's not really adoption, and they don't get to visit.

Then there are the zoos that raise money from the public to support captive breeding programmes of endangered animals. Probably the best recent examples are the Smithsonian and Edinburgh Zoo's attempts to help giant pandas to help themselves.[2] These rather feckless animals have ambled slowly up an evolutionary cul-de-sac by not only having a small period in the year when the female is fertile, but also by having little enthusiasm to do the necessary when the time is right. Typically, a female panda is fertile for between 24 and 36 hours each year, and so zoos have come to think they can manage all this better than the animals do themselves. In the 1980s, the Chinese government lent out its pandas as part of its foreign policy, but now it asks for large sums. Zoos round the world once paid $1 million for pairs of animals, though the price has fallen lately. Zoo strategy is to hire the pandas and then charge a willing public to cover the cost. Henry Nichols, writing in The Guardian, reckons it cost Edinburgh Zoo £250,000 to construct a new enclosure, and it will be paying China around £640,000 a year for 10 years to hire its pandas. It also costs £70,000 a year for their food, bamboo, which is imported from Amsterdam. Add to this the salaries of keepers and you get somewhere near the full cost.

Although people never seem to tire of looking at giant pandas, it's not obvious that exposing the animals to public gaze provides whatever level of intimacy that they need to get together. In Edinburgh, after four years, there was no sign of junior pandas, despite repeated attempts at consensual sex and artificial insemination. These days, tweets from #pandacam are an essential part of the visitor experience. OneKind, an animal protection charity, criticises the use of artificial insemination, saying that pandas should be left in peace as it is misguided to breed yet more captive animals in zoos when they will never return to the wild; nor, it says, will this result in improved protection for the wild population in their native habitat.[3]

Writing about the Smithsonian panda Bao Bao, which actually managed to give birth in 2015 thanks to two sperm donors, artificial insemination and numerous midwives, The Economist said that the panda:

> should fit in well in Washington: she costs a fortune, has no useful skills and is always on TV. On the plus side, she is better-looking than any member of Congress – and unlikely ever to be involved in a sex scandal.[4]

It's important to say, however, that there are positive stories about species that were almost extinct in the wild that have been rescued. Californian condor numbers dramatically declined in the twentieth century because of poaching and habitat destruction. The remaining 22 wild condors were captured in 1987 and a captive breeding programme started. Starting in 1991, condors have been released, and there are now about 400 of them living wild or in captivity. The WWF website[5] is a good source of information on such schemes.

Notes

1 Wikipedia is a good source of information on endangered species. Web link: en.wikipedia.org/wiki/Endangered_species.

2 Nicholls, H. (2011). What price captive pandas? *The Guardian*, 2 December. Web link: ow.ly/niqA308QNkA.

3 Packham, C. (2015). Chris Packham on zoos. *BBC Wildlife Magazine*, 3 June. Web link: ow.ly/3v4j308QNvQ.

4 *The Economist* (2014). Pandanomics: costly, bumbling Washington has a perfect mascot. 18 January. Web link: ow.ly/2guz308QNKX.

5 The WWF website. Web link: wwf.org.uk.

27

THE GREENHOUSE EFFECT

The Earth is warmer than it ought to be, given its distance from the Sun, and this is because of the gases in the atmosphere surrounding us. These are said to hold the heat, so that the atmosphere acts rather like a greenhouse. It is a natural effect, and the Earth is (on average) about 32°C warmer than it otherwise would be. This means that we humans can live here reasonably comfortably. The way it works, however, is not quite like the way a greenhouse does, but it's a memorable image, so people use it.

The atmosphere contains nitrogen (about 78%), oxygen (about 21%), argon (about 1%), carbon dioxide (about 0.04%), and a lot of other gases in much smaller proportions. Some of these are pollutants; for example, carbon monoxide, methane, nitrous oxide, and the many chlorofluorocarbon (CFC) gases that were once used in fridges and aerosol cans before they were banned through the Montreal Protocol. The global warming potential (GWP) of all greenhouse gases is measured against that of CO_2. GWP is based on the heat-reflecting properties of the gas and the length of time it remains in the atmosphere. Over a 100-year period, CO_2 has a GWP of 1, compared with 25 for methane and 298 for nitrous oxide. The GWP of CFC gases are in the tens of thousands.

There is water vapour in the atmosphere as well, which is also a greenhouse gas, but it tends not to be included in the list. One of the gases, ozone, found in the high-level stratosphere, has a vital role in protecting us against radiation from space. Carbon dioxide (CO_2) and oxygen also have vital roles; all living things need oxygen for respiration and the production of the energy we need to live, and plants need carbon dioxide for photosynthesis and growth. By a neat twist, CO_2 is a by-product of respiration, and oxygen is a by-product of photosynthesis. Forests, because they generate so much of the oxygen that we need through photosynthesis, are sometimes referred to as the lungs of the world. This is another memorable image, but is quite misleading, as lungs take air in and out of the body, but there is always more carbon dioxide going out than going in.

Carbon dioxide is very effective at trapping heat, and so is said to be a potent greenhouse gas. In the past, the natural processes that remove CO_2 from the atmosphere, such as photosynthesis and absorption by the oceans, more or less balanced out the amount released by respiration and the decomposition of organic matter on the land and in the oceans. According to NASA, this two-way exchange of CO_2 involves around 220 billion tonnes of gas a year. Since the Industrial Revolution, however, the amount of carbon dioxide produced by industry, engines, heating and power sources has steadily risen, causing the atmosphere to warm up and cause what is known as the enhanced greenhouse effect, giving rise to what we now call global warming. NASA estimates that about 3% of the natural CO_2 exchange is additional gas released from fossil fuels each year. That may not sound much, but the evidence is that it has upset a natural balance.

These human-produced greenhouse gas emissions have increased since the start of the Industrial Revolution around 1750. They were driven by economic growth and population increase. There was a huge increase over the last years of the twentieth century. For example, in the 1950s, we were releasing around 6 billion tonnes of CO_2 into the atmosphere every year, but in 2014 we released over 35 billion. Another way of seeing this change is to look at how the proportion of CO_2 in the atmosphere has changed. In pre-industrial times (the 1750s), the amount of CO_2 in the atmosphere was 0.028%. In 1950, it was 0.031%. In September 2016, it was 0.040%, with about half of the extra CO_2 coming in the last 40 years. If you go to NASA's Earth's Vital Signs website, you'll be able to see today's concentration, as well as historical data.

Industrial development has led to atmospheric concentrations of carbon dioxide, methane and nitrous oxide that are unprecedented over the last 800,000 years. The UN's International Panel on Climate Change (IPCC) says that their effects have been detected throughout the climate system and are extremely likely to have been the dominant cause of the observed warming since the mid-twentieth century, and the climate change that has taken place.[1]

Although we've been pumping all this CO_2 into the atmosphere, not all of it stays there. Some finds its way into what are known as sinks. The oceans are a big CO_2 sink, and they are becoming more acidic as a result. Unsurprisingly, photosynthesis is another sink, and one result of the extra CO_2 in the atmosphere is an increase in plant growth, which includes plants surviving in places where they don't usually grow. NASA says that towards the end of the twentieth century, about 50% of the CO_2 generated was absorbed by this plant growth, and that by 2015 it had risen to 60%.[2,3] NASA also says that in the 26 years between 1982 and 2009, an extra 18 million square kilometres became covered in vegetation – that's about twice the area of the US. This is doubly good news. Not only is some of the extra CO_2 we're generating being reabsorbed, but also there is more vegetation for animals to eat, which will also result in more animals in areas of the world where they once might have been scarce. Sadly, this won't go on being such positive news forever. First, this will only slow climate change, not stop or reverse it. Next, plants do not live by CO_2 alone, but need water, and nutrients from the soil, and these might be in short supply in the areas where most of the vegetation growth is occurring. Then, increasing the land coverage by plants will make the Earth better at absorbing (and worse at reflecting) radiation from the Sun. This always warms the Earth.

The greenhouse effect was discovered in 1856 by the American scientist Eunice Foote.[4] She showed that gases in the atmosphere were affected by the Sun's radiation in different ways, and of all the gases she tested it was CO_2 that trapped the most heat. In 1859, UK physicist John Tyndall independently demonstrated the same effects, and he usually gets the credit for making the link. This was because he published his results, including data on how the absorption of radiant heat differed from gas to gas. Foote was ahead of her time. Although her work was not published, and she wasn't even allowed to present her findings to fellow scientists as no woman could do that in the US at the time, it's reported that she also speculated about whether changes in the amount of carbon dioxide in the air might lead to a warming of the Earth. Rather inevitably, only one of these two scientists has a university climate change research centre named after them.[5]

Notes

1 The IPPC has a lot of data on the greenhouse effect. Web link: ipcc.ch.
2 The NASA website has a good account of the Earth/ocean system. Web link: ow.ly/GSmC3o8QvvZ.
3 The NASA Earth's Vital Signs website shows the latest concentration of carbon dioxide in the atmosphere. Web link: ow.ly/6qSO3o8Qv83.
4 Details of the work of Eunice Foote can be found online. Web link: ow.ly/IvRF3o8QuEp.
5 The University of East Anglia's Tyndall Centre is named after John Tyndall. Web link: tyndall.ac.uk.

28

VALUING THE ENVIRONMENT

The phrase the environment can have very specific meanings or can mean absolutely everything that surrounds us. We can add a wide range of adjectives to it to create phrases such as economic environment, built environment, dog-friendly environment, etc., and we talk about the extent to which people are affected by their environment. However, as we saw in our chapter on nature, when we talk about the environment we are saying something about the meaning that our surrounding world has for us. The following anecdote illustrates this well.

During a research project on education for sustainability in Africa in the 1990s, an 80-year-old Malian woman, Mrs Samake, was asked what she understood by the term environment. She replied: "We do not talk about environment, we talk about what is important to us: gardening, useful trees and culture."[1] Mrs Samake would have had no truck with the idea of environment as mere surroundings. To her, the environment clearly is a part of what makes us who we are — and we are a part of it. This is something that Western thinking seemed to lose during the time of the Enlightenment, when, for example, the seventeenth-century philosopher

René Descartes famously drew a distinction between humanity on one side and nature on the other. This separation is often blamed for our disastrous environmental record during our pursuit of industrial development.

The Enlightenment ushered in a scientific approach of analysing individual components of systems in isolation; this has served us well with numerous insights and innovations. The downsides, in terms of an atomised and utilitarian view of the world, not to mention pollution and habitat loss, were already apparent in the nineteenth century when the Romantic movement sought to value the environment for its own sake. By the twentieth century, the writer Aldo Leopold argued for an extension of our ethical codes to include animals, water, soil and air, what he referred to as the biotic community.[2] Leopold proposed a land ethic that involved a view of the natural world not dissimilar from that of many pre-industrial societies: "a land ethic changes the role of Homo sapiens from conqueror of the land community to plain member and citizen of it. It implies respect for his (sic) fellow-members, and also respect for the community as such." However, in the face of competing economic interests, citing the connectedness of the environment in general terms is not good enough. Defenders of the environment need to provide at least one of two things:

(i) A readily understandable scientific explanation of its ecological value, both in terms of the role it plays as part of the wider web of life and its current vulnerability.

(ii) A plausible calculation of the economic value of the environmental goods and services that it provides to humans. This might include an estimate of the costs that would be incurred in attempting to replace it.

The second approach has gained popularity with economists, but it's not a straightforward business. In the 1990s, IUCN produced estimates of the economic value of the environment and proposed solutions such as debt-for-nature, where poorer countries were paid to retain high levels of biodiversity. This pragmatic attempt at valuation has been repeated at the national level; for example, the UK National Ecosystem Assessment, launched in 2011, reported that the country's parks, lakes, forests and wildlife were worth billions of pounds to the economy. Such work is often criticised for reducing the environment to mere money. After all, if our natural capital really were converted into cash, there would be little hope for life on Earth, no matter how fairly the money was distributed.

Such calculations are also criticised for being too human-centred or anthropocentric. After all, who are we to place a value on aspects of the environment that we do not make use of ourselves? Isn't that like playing god? And what about emotional attachment? What price would we put on something that is loved for simply being there?

In their study "What is the value of Rangitoto Island?," Dan Vadnjal and Martin O'Connor asked residents of Auckland, New Zealand, how much they would be willing to pay to protect a nearby unspoilt volcanic island.[3] The majority of answers did not fall within the sphere of conventional economics; instead, people suggested paying nothing or an infinite amount. This is because it proved impossible to reduce something as special as a pristine environment to money. As one resident put it: "There are things, dimensions of life, that are beyond money. With money you can buy a car, and it rusts . . . A car and Rangitoto do not exchange. So what is the value of Rangitoto? It exists." People also expressed dismay at the thought of paying for something that they thought was already theirs. This may seem an extreme case, but strikingly similar sentiments were expressed in the UK when the government attempted to sell off publicly owned forests in 2010. However challenging it is to grasp the complex nature of the environment, we are certainly not ready to give it up wherever we have a sense of connection.

The work of environmental education has for years sought to develop our environmental literacy, in other words, our knowledge, skills and dispositions in relation to the environment. This often focuses on giving young people direct experiences out of the classroom to develop pro-environmental sentiment as well as understanding. Equally crucial, however, is developing a willingness to engage in civic life, to participate in decisions that will have far-reaching environmental impacts. We live in a time of increasing concern with young people's isolation from nature, to the point of there being an identified condition of nature deficit disorder.[4] We also see the arrival of the screenager, a term coined by the author Douglas Rushkoff to describe young people who spend large amounts of time on computers, mobile phones and games consoles.[5] The significance of our living in this virtual environment is not yet fully understood.

Our modern understanding of the world is often coloured by familiarity with machines, but however complicated these are, they provide poor analogies for the complex, non-linear way that the environment appears to work. That said, our participation within the myriad connections of the Internet does reflect something of the complexity of natural systems, with their intricate web of connections that links unfamiliar systems to our own, and ultimately to the global environment.

It seems ironic that this virtual environment, which apparently isolates us from the physical world, may yet provide the best analogy we have for the complexity of the world. New ethics or protocols that are even now emerging among Internet users may yet guide us in defining a new land ethic for our time. Gradually, we are (re)learning that we are part of everybody else's environment and they are a part of ours – we just need to be mindful of the fact that this extends well beyond humans to include the biophysical world.

As our chapter on nature suggests, the environment can be seen to be as much a part of us as apart from us. An understanding of complexity tells us that what you choose to do or not do in the environment is making some sort of difference even as you read this. Ultimately, the way that we choose to value the environment will probably say a lot about the way we value each other.

Notes

1 Vare, P. (1998). ECoSA: a report on a pan-African environmental education survey. *Environmental Education Research*, 4(1), 5–24.
2 Leopold, A. (1949). *Leopold's Land Ethic*. Baraboo, WI: The Aldo Leopold Foundation.
3 Vadnjal, D. & O'Connor, M. (1994). What is the value of Rangitoto Island? *Environmental Values*, 3(4), 396–380.
4 Louv, R. (2005). *Last Child in the Woods: Saving Our Children from Nature-Deficit Disorder*. Chapel Hill, NC: Algonquin Books.
5 Rushkoff, D. (2006). *ScreenAgers: Lessons in Chaos from Digital Kids*. New York: The Hampton Press.

29

SUSTAINABLE DEVELOPMENT

Gro Harlem Brundtland was Norway's prime minister for three periods between 1981 and 1996. In 1983, she was invited by the United Nations to chair the World Commission on Environment and Development (WCED). In time, this became known as the Brundtland Commission. Its report, *Our Common Future*, developed the idea of sustainable development, which in turn became a key focus of the 1992 Earth Summit, and the action plan that was developed from this, Agenda 21.[1,2]

In *Our Common Future*, the text saying what sustainable development means, begins: "It is development that meets the needs of the present without compromising the ability of future generations to meet their own needs." Although the report went on to elaborate the idea, it is only this sentence that is usually quoted, even by people and organisations who know that it's much more complex than this suggests.

However, the sentence does say two important things:

(i) It sets out a clear moral commitment to future generations, with a warning to those of us on the Earth at the moment. It says: don't mess with the planet's ability to support life because others – those unborn

and unbegotten – will need that too. This contains the idea of stewardship: that we are only looking after the planet for those who follow,
and we have a duty to pass it on in good condition.

(ii) It also sets out a moral commitment to everyone who is on the Earth
now. Although this is more implicit, it is obvious that you cannot
speak of meeting "the needs of the present" unless you include everyone. Indeed, the very next sentence that follows this oft-quoted line
stresses the needs of the Earth's poorest people.

It's clear from this that sustainable development is about the ability of all of
humanity, now and in the future, to live, as Amartya Sen put it, "a life they
have reason to value."[3] The implication of having sustainable development
as a policy goal is that there needs to be a focus on human wellbeing and on
a well-functioning biosphere in order to make widespread and enduring
human fulfilment a possibility.

All this is fine, up to a point. The problem is the word needs. You can see
why Brundtland used needs; it has a long history in relation to ideas about
human and social wellbeing. It also carries within it the idea of a minimum
entitlement to the fundamental essentials of a good life.

Another well-known phrase with needs in it springs to mind: "From
each according to his ability, to each according to his needs." Although
Karl Marx is popularly thought of as the author of this socialist slogan, it
was probably first used by Louis Banc in 1851. Its origins may lie in the
writings of the French utopian Étienne-Gabriel Morelly, who proposed a
Code of Nature in 1755 which included the idea that "Nothing in society
will belong to anyone, either as a personal possession or as capital goods,
except the things for which the person has immediate use, for either his
needs, his pleasures, or his daily work."

Some say that the idea can also be found in the Bible in the Acts of the
Apostles (4:33–35):

And with great power gave the apostles witness of the resurrection of
the Lord Jesus: and great grace was upon them all. Neither was there
any among them that lacked: for as many as were possessors of lands or
houses sold them, and brought the prices of the things that were sold.
And laid them down at the apostles' feet: and distribution was made
unto every man according as he had need.

But the problem is that we in the developed West, and everyone else who has plenty of money, are not really concerned about needs. Rather, our focus is almost always on wants. That is, all the things that we might get our hands on in order, or so we think, to enrich our lives. Such wants are not just material things (stuff), but are experiences as well.

In a modern capitalist economy, we are not really trusted to think for ourselves about what our wants (or needs) might be, so the advertising industry is there to help us along. Whether on TV, in newspapers and magazines, and increasingly online, we are constantly reminded of what we don't have and encouraged to spend money on it. Indeed, modern economies need people to spend money on their wants in order to provide employment and tax revenues. If we only spent our money on our needs, the results would probably be calamitous. Happily for the global economy, this is unlikely to catch on, especially among the emerging wealthy middle classes in so-called developing countries.

All this implies that sustainable development has to take wants into account if it is to be taken seriously. Perhaps it is ideas such as the circular economy that may help here, with its key notion of separating economic growth from resource use, and therefore need.

Notes

1 World Commission on Environment and Development (WCED) (1987). *Our Common Future*. Oxford: Oxford University Press.
2 *Our Common Future* can be found online. Web link: un-documents.net/our-common-future.pdf.
3 Sen, A. (1999). *Development as Freedom*. Oxford: Oxford University Press.

30

DEVELOPMENT

Development is a slippery term with multiple meanings, especially when it comes to talking about the means by which we compare nations, claiming some to be developed and others less or underdeveloped. The idea of development aid is widely accepted, after all, and who would be against richer countries helping poorer ones? If only it were that benign. In *The Development Dictionary*, Wolfgang Sachs traces the current use of the word development to US President Harry Truman's 1949 inauguration speech.[1,2] Truman defined the largest part of the world as underdeveloped, as if all countries are on a single path with frontrunners and laggards, which has the sense of making *them* more like *us* – whether they like it or not.

A distinction can be made between *economic development*, a measure of a country's wealth, and *human development*, which is a broader measure of the wellbeing enjoyed by the citizens of a country; this may include health, education, political rights and income. Wealth is normally measured by either gross domestic product (GDP), that is, the monetary value of all the goods and services produced in a country, or gross national product (GNP), which is a country's GDP plus inward foreign investments.

These are crude measures that tell us nothing about how a nation's wealth is used or distributed. Pockets of severe poverty exist in apparently rich countries (and vice versa), while many factors can lower the quality of life of some citizens. To measure equality within a country, a statistical measure of distribution is often used called the Gini coefficient (named after the Italian statistician who devised it). A perfectly equal distribution would give a Gini score of 0, while a score of 100 would suggest that one person had all the wealth. Typically, Scandinavian countries have the most equal distribution, with Gini scores in the low 20s, while southern African countries are the most unequal, scoring in the 60s. In 2010, the UK score was 32.3, and the US's 45.

Rather than relying on wealth alone, the United Nations Development Programme (UNDP) has ranked countries according to a composite measure called the Human Development Index (HDI). This considers three sets of indicators covering health, education and standard of living. The agreed goal for the HDI is:

- life expectancy of 85 years;
- adult literacy and educational enrolments of 100%; and
- GDP per person equivalent to US$40,000 in purchasing power parity (this takes into account a country's cost of living and inflation rate).

The actual HDI indicators for a country are expressed as decimal shares of this ideal.[3]

This still doesn't tell us much about the relative importance of the different components or how these are impacting on the poorest. So UNDP, with the help of Oxfam, has established a more sophisticated measure called the Multidimensional Poverty Index (MPI). This has 10 indicators, including nutrition, sanitation, level of schooling, and access to cooking facilities and electricity, and elements that help to explain how people feel poor.

Another way of measuring development is Bhutan's Gross National Happiness Index (GNH).[4] This has four pillars (political, economic, cultural and environmental) that are subdivided into nine domains that include psychological wellbeing, health, education, governance, culture and ecological diversity. GNH scores are based on a survey with 148 questions that researchers discuss with about 8,000 residents from across the country every five years. Between 2010 and 2015, there was a slight increase in

GNH despite a dip in governance scores, which was attributed to people then being less euphoric than they'd been in 2010 following a recently acquired democratic constitution. However, an aggregate score for happiness, across thousands of people, may not be very meaningful when the people themselves don't get to choose the relative importance of their own responses. We will return to this issue below.

Happiness is all very well, but if we're concerned that the very fabric of life on Earth is unravelling, we'll need a stronger emphasis on ecology and biodiversity. This is something the Happy Planet Index (HPI) provides.[5] It compares how efficiently citizens use natural resources in different countries in order to achieve wellbeing. A simplified version of the HPI equation is:

$$HPI \approx \frac{(\text{Life expectancy x Experienced wellbeing}) \times \text{Inequality of outcomes}}{\text{Ecological footprint}}$$

Figure 30.1   A simplified Happy Planet Index equation

Source: New Economics Foundation

This incorporates inequality as well as a self-reported measure of wellbeing and a hard statistic (life expectancy) to indicate health. The ecological footprint takes into account all goods and services used rather than produced in a given country, so it is a measure of consumption, not production.

Although some might be tempted to claim that all these different measures are now well covered by the Sustainable Development Goals (SDGs), these are extremely wide-ranging, and therefore even less likely to achieve a meaningful score than, say, Gross National Happiness.

A former Chief Economist with the World Bank, Joseph Stiglitz, makes the point that economically developed and less developed countries are actually separated by a gap in knowledge rather than the amount of resources available.[6] After all, improved standards of living have resulted from technological advances, not simply from the accumulation of wealth. For Stiglitz, it is learning – and more importantly, learning to learn – that will close this gap. In calling for a learning society, he claims that neoliberal policies imposed on developing countries by organisations such as the

International Monetary Fund (IMF) have hindered rather than helped the creation of such a society.

One feature shared among the measures discussed here is that they have been devised by experts – often in richer countries – and imposed on others; as such, these inevitably represent the values of those who developed them. That said, we are getting smarter with these calibrations. For example, the website of the Your Better Life index (created by the OECD) allows you to choose the relative importance of different measures so you can see how countries are ranked according to the criteria you decide to emphasise – although out of necessity, the criteria have been pre-selected from available data sets.[7]

An alternative to using these expert-generated measurements is to look at people's ability to achieve what they want. For Nobel laureate Armartya Sen, human development rests on people having the freedom to lead lives that they have reason to value.[8] This is quite different from having your life measured against the things that experts have reason to value on your behalf. Sen calls this a capabilities approach, and, like Stiglitz, he prefers to see development in terms of people's degree of self-determination, as well as their skills and knowledge. The focus here is people's freedom to learn and make informed choices. If this were supported by freely available education that had a solid grasp of ecology at its core, then a learning society might indeed become an enduring measure of sustainable development.

Notes

1 Sachs, W. (1992). *The Development Dictionary: A Guide to Knowledge as Power.* London: Zed Books.

2 Arrow, J.K. (1962). The economic implications of learning by doing. *The Review of Economic Studies*, 29(3), 155–173. This inspired Joseph Stiglitz's view of a learning society.

3 Oxford Poverty and Human Development Initiative (2016). The Global Multidimensional Poverty Index. Web link: ow.ly/byoM308Son2.

4 Bhutan's 2015 Gross National Happiness Index is on the Centre for Bhutan Studies website. Web link: ow.ly/DQ6O308RZlx.

5 New Economics Foundation (2016). Happy Planet Index. Web link: happy planetindex.org.

6 Stiglitz, J. & Greenwald, B. (2015). *Creating a Learning Society: A New Approach to Growth, Development, and Social Progress.* New York: Columbia University Press.

7 OECD (2016). Your Better Life Index (2016). Web link: oecdbetterlifeindex.org.

8 Sen, A. (1999). *Development as Freedom.* Oxford: Oxford University Press.

31

THE FOUR CAPITALS

Sustainable development has social, environmental and economic dimensions to it, and it demands positive change in all three of these at the same time without trade-offs between them. In other words, economic growth that trashes the planet is not sustainable; nor is environmental protection that makes people destitute, or social change that wrecks an economy. This means that sustainable development can be almost as difficult to talk about as it is to carry out.

One way of thinking about it is in terms of the assets we draw on, and the technologies and institutions we use, to improve how we live: our civilisation and our wellbeing. Herman Daly and Donella Meadows described this in terms of four capitals: natural, built, human and social.[1]

Our version of how they thought about this is set out below. This shows the relationship between our political economy and nature, and how all life, and the economic transactions that underpin it, are ultimately supported from within the biosphere. It also sets out the ends to which these are put. In this view, what really ought to matter to us is beyond the stuff we buy and surround ourselves with, and how we communicate. What should matter is beyond even our health and wealth. In this view, all these are merely the means to

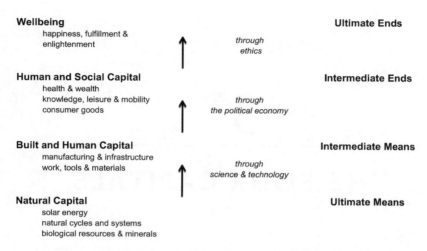

Figure 31.1 The relationship between the political economy and the biosphere

Source: Donella Meadows

wellbeing wherein lies true happiness, human fulfilment and enlightenment. Clearly, not everyone will share this view of what life is all about.

However, you don't have to share this view of wellbeing to see the value in its way of thinking about the world, as there are important truths within the model about how we live.

Fundamentally, supporting absolutely everything are what Herman Daly called the ultimate means on which all life and economic and social transactions are based: natural capital. This is energy from the Sun, the Earth's natural materials cycles and systems (for example, the carbon and nitrogen cycles, and ocean and atmospheric currents), all living things on the planet with their ecosystems and habitats and the genetic information they contain (including we humans), and all the mineral resources found in the Earth's atmosphere, oceans and land. Everything we do and make and know is based on this capital stock. This was not created by clever humans. Rather, it is our legacy from the past. Donella Meadows says that it is the heritage that we were born into, although we rarely treat it as such.

The intermediate means are what we use to process and convert natural capital to useful things. These means include tools, machines and devices of all kinds, the factories, offices and systems in which they are used, and the skilled labour that people provide that is needed for all these. As such, these also include schools, hospitals, transport systems, and much more. They constitute our built and human capital. These intermediate means set the limits for how wide-ranging an economy will be. Such means are necessary if we are to realise what we want from life, but they are not sufficient to do this. Managing these intermediate means is the concern of economics, social systems and politics; that is, the political economy. We use this to value, distribute and maintain resources across the societies in which we live.

Intermediate ends are different. These are what governments promise us, and what economies are expected to deliver, if we work hard enough: health, wealth, knowledge, leisure, transportation, and consumer goods and material things (all that stuff we surround ourselves with). Although this is what everyone says they want, satisfaction is not always guaranteed. This is the case even when our wants are delivered, as is shown when intermediate ends are in abundance but people feel their lives are empty.

That is because intermediate ends are not ends in themselves, but instruments to achieve something yet higher – the ultimate ends of happiness and human fulfilment. Or so philosophers tell us. A recent World Bank survey included this quote: "I like money and nice things, but it's not money that makes me happy. It's people." It is now widely recognised that where you are socially isolated, that is, you don't have other people in your life, as many older people now don't, then your life is poor indeed – no matter how much cash you have.

Note

1 Donella Meadows discusses the drawbacks of the four capitals model online. Web link: nssd.net/pdf/Donella.pdf.

32

HARMONY

It might seem odd that there's a chapter in this book on harmony, and before you read on, you might ask yourself what springs to mind when that word is mentioned.

If you're at all musical, then you might think of the pleasing sounds that come from a certain combination of notes, or even of the harmonics that are so important to the richness of the sounds we hear. Or, thinking of St Francis of Assisi, it might be the sort of harmony that seeks to displace discord, or the harmony we associate with tranquility and peace when we speak of living in harmony with our neighbours, or of getting along. You might have also thought of Harmonia, the Greek goddess of peace and concord. Maybe eHarmony, the dating website, came to mind, or Harmony the rapper, who's a hip-hop emcee.

Then again, you might know that in 2009, the United Nations[1] proclaimed 22 April as International Mother Earth Day. In doing so, it acknowledged that the Earth and its ecosystems are "our home," and expressed its conviction that it is "necessary to promote harmony with nature in order to achieve a just balance among the economic, social and environmental needs of present and future generations." Or perhaps you know that in 2010, the Prince of Wales wrote a book (with Tony Juniper and Ian Skelly) with the title *Harmony: A New Way of Looking at Our World*, with a focus on the world's environment and its problems.[2] It's for these last two reasons that we are interested in the idea.

The UN said that the world has been slow to respond to the damage that human activities are causing to the planet, and that the purpose of International Mother Earth Day was to promote a view of the Earth as the entity that sustains all living things found in nature; it was to honour the Earth as a whole and our place within it.[3] Confusingly, the UN had, in the 1970s, already designated 22 March as Earth Day. The differences between the two are subtle.

The Prince and his co-authors say that *Harmony* is a practical guide to what we have lost in the modern world, why we have lost it, and how we might rediscover it. It's a philosophical blueprint for the "more balanced, sustainable world" that we'll have to create if we're all to live well in the world. In this sense, it's rather like the Earth Charter. *Harmony* looks at how many of the world's challenges can be traced to how we have abandoned a "sense of balance and proportion," and it illustrates how many of the practices of modern life have put us at odds with the rest of nature. It sets out to show how this imbalance influences our lives for the worse. It tells the story of what the authors see as our disconnection from nature and what this has contributed to the greatest crisis in human history. It argues that if we seek balance in our actions, this will return us to a more considered, secure, comfortable and less polluted world. The key idea here, and in many of the other meanings of harmony, is mutually beneficial coexistence.

As we have seen, the word balance features a lot in the Prince's book, but balance and harmony are not the same thing. Balance is one of those words that everyone says they understand, when perhaps their grasp is shaky. What image comes to mind when balance is mentioned? Is it a see-saw? Or a pair of old-fashioned weighing scales where you put the weights on one side and the thing to be weighed on the other? We need a (perhaps intuitive) understanding of Newton's gravitational theory to make sense of both of these.

But the idea of a balanced diet is quite different, and is closer to the sort of balance that the Prince of Wales is writing about. Understanding the idea of a balanced diet requires a theory of nutrition that allows you to know what to include in a diet, and in what proportion. So what sort of theory do you need if you are to promote "harmony with nature" in order to achieve the UN's "just balance among the economic, social and environmental needs of present and future generations"? The UN doesn't say.

And are we interested in harmony with nature, or within it? The first sees humans separate from nature; the second as part of it. This second idea, harmony within nature, might be useful as a political philosophy that seeks to protect the biosphere from the hubris of humans, but it does not describe nature itself, where you see inequality, competition, violence and death playing out all the time. There may be lots of examples within nature of mutually beneficial coexistence, and we might want to draw on this to inform our own philosophy and politics, but you can never fully escape the other metaphor of nature's being "red in tooth and claw."

In a *Guardian* review of *Harmony*, Terry Eagleton argued that the book's unifying thread is "the need to abandon a soulless modernity for a traditional spirituality."[4] In other words, this isn't a question of acting or behaving differently, but about thinking differently; that the biggest problem we have is how we look at, and see, the world. *Harmony* argues that we should look at it like we used to, when we had a world view that saw humanity as a part of nature. Now, modernism and the Enlightenment have separated us from nature. While *Harmony* argues that solutions to our various crises (climate change, poverty, species extinctions, etc.) lie in regaining that balance with the world around us, what it doesn't do is tell us how. That's because it cannot.

All this takes us back to our default position of seeing harmony lying in our seeking to establish mutually beneficial coexistence, not only with other humans, but with the rest of nature. We might possibly do this through nature-friendly policies, less exploitation of natural resources and wild places, living with a much smaller ecological footprint, and doing what we can to lessen our impact on other species, habitats and ecosystems.

We might try to have a kind of binding contract with nature to bring all this about, and might envisage a United Species Organisation to police it. But this would be a one-sided negotiation. We might want to sign up, but why would nature, even were it possible? Nature is supremely indifferent to us, to human hopes and dreams, vanities and foibles, and hubris. Fundamentally, nature is not on our side, and we might remember that whenever we hear anyone speaking of harmony within nature as though it had much meaning.

Notes

1 United Nations Harmony with Nature. Web link: harmonywithnatureun.org/index.html

2 HRH Prince of Wales, Juniper, T. & Skelly, I. (2010). *Harmony: A New Way of Looking at Our World*. London: HarperCollins.

3 International Mother Earth Day. Web link: ow.ly/5nNL308S1Fp.

4 Eagleton, T. (2010). *Harmony: A New Way of Looking at Our World by HRH The Prince of Wales – Review*, 6 November. Web link: ow.ly/zcNO308S1m5.

33

SYSTEMS AND SYSTEMS THINKING

The word system is in common use with a range of meanings. You often come across it as a two-word phrase: digestive system is a good example. Having mentioned this, others will probably come to mind, such as solar system, legal system, transport system, nervous system and weather system. There are dozens like these, including ecosystem. Although the examples listed are different, they all have some common features: they contain separate parts that are connected in some way; there is a boundary of some kind; they all involve some sort of movement or flow, or inputs and outputs; and, if you want to understand how it all works, you have to think of the parts in relation to the whole (not just the parts separately). For example, the digestive system is called a system because all the bits – the stomach, liver, salivary ducts, small intestine, gallbladder, etc. – are not only all connected physically, but they also influence each other.

All this applies to time as well as space, and to relationships between ideas and people, as well as between physical objects. Typically, systems are characterised by five features:

Table 33.1 Five features that characterise systems

System features	Examples
A dynamic and complex whole made up of connected parts	Bodies and engines
Systems can be within other systems (and parts can be systems)	Memory chips in computers
Energy, material and information can flow between the parts of a system	Between oceans and atmosphere in our climate system
Systems can be completely open to the outside or closed	A lake (open) or the inside of a perfectly insulated box (closed)
Systems can be closed to some things, but open to others	The Earth is open to energy and information, but closed (mostly) to material

Source: Linking Thinking

Seeing the world like this is in contrast to the way of thinking ascribed to the seventeenth-century French philosopher René Descartes, whose ideas dominated scientific thought and activity up to the mid-twentieth century. His reductionist approach led us to think of parts and not the whole in a rather mechanistic way. The most glaring example of this, if you think about it, is how we conceive of mind and body as separate.

In a reaction to the problems reductionist thinking causes, the twentieth century saw a growing focus on not only thinking about systems, but also of thinking in systems terms. This is an approach to problem-solving that attempts to balance wholes and parts. This is systems thinking, and is also described as holistic, relational or ecological thinking.[1]

This systems thinking is quite different (and harder) than thinking about systems, but some people find it straightforward. Perhaps you are a natural systems thinker. Do you, for example, look at the context of an event (the big picture) as well as the event itself? Do you look for connections, thinking about this as widely as possible? Do you look for explanations, other than the obvious ones? Do you think about the longer term as well as the immediate?

If you said "yes," even if only some of the time, there is something of the systems thinker about you. If so, you might tend to assume that:

- the whole is usually more than the sum of parts, and we cannot really understand something just by looking at the parts;
- most processes are not linear, with clear beginning and end points;
- there isn't necessarily a solution to every problem;
- we can't always best understand things through a rational approach, downplaying intuition;
- people are not separate from nature, nor are facts separable from values; and
- objectivity is not always possible if we're fully to understand issues.

Maybe you're a bit of both; it would be surprising if that were not the case. Maybe the systems thinker is trying to break free from the hold that linear thinking has on you. That would also be unsurprising given how much we are influenced by media, much of which is always looking for simple cause and effect explanations and avoiding complex, nuanced arguments. Some say that the education that most of us have had has contributed to this problem. This contrasts with how we all begin life as babies when, in our early explorations of the world, we look for connections and relationships, and for similarities and differences. As Peter Senge says: "It appears we have latent skills as system thinkers that are undeveloped, even repressed by formal education in linear thinking."[2]

As a culture, we tend to be good at analysing things, but not so good at synthesising novel solutions; good at categorising things, but not so good at seeing commonalities across categories; good at seeing detail, but not so good at identifying pattern; and good at looking at issues separately, but not so good at recognising relationships between them.

Consider air pollution in our cities. We know that it's caused by the emission oxides of carbon and nitrogen, small particles, and other substances from vehicle exhausts, and we can measure the amount with great precision, showing how it changes over time and from place to place. But is this a problem caused by poor engineering? For example, if we made engines produce fewer pollutants, would that solve the problem? Is it, in other words, just a technology issue? Or are there just too many vehicles, too many journeys or too many bad drivers? If so, what are the causes? Does part of the problem lie in how we plan our living spaces, or perhaps our absolute right to own and drive a car more or less wherever we like?

Systems thinking is increasingly being used to tackle a wide variety of fields such as computing, engineering, epidemiology, information science, health, manufacture, management, development and the environment. Its supporters argue that we need this way of thinking as the world we now live in is quite different from before in three significant ways:

- *Complexity*: Environmental and human systems are now complex in the way described above, and are made so through globalisation and advances in communications.
- *Uncertainty*: Although we live in cultures that demand certainty, prediction and control, in such complex contexts it's impossible to determine or predict outcomes with precision. Because of this, we will all need to become more tolerant of ambiguity and uncertainty.
- *Unsustainability*: Complexity, uncertainty and unpredictability create doubts about the sustainability of social and economic systems. Meanwhile, the biosphere is itself looking increasingly unsustainable.

Generally speaking, it is the young who, rather instinctively, have a better chance of understanding all this than their parents and grandparents. We find that reassuring. Education, too, has suffered from a disconnected view of the world. Specialisation and efficiency have only ensured that we know more and more about less and less. Knowledge is emphasised over wisdom, and what the Prince of Wales calls a "whole-istic" education that connects children with nature and their heritage may be almost lost.[3]

Notes

1 Linking Thinking sets out perspectives on thinking and learning for sustainability. It was commissioned by WWF Scotland and written by a team led by Stephen Sterling. Web link: ow.ly/OozW308LILb.
2 Senge, P. (2006). *The Fifth Discipline: The Art and Practice of the Learning Organisation*. London: Random House.
3 HRH Prince of Wales, Juniper, T. & Skelly, I. (2010). *Harmony: A New Way of Looking at Our World*. London: HarperCollins.

34

FRAMES AND FRAMING

Mental frames are so much part of our understanding of the world that most of us are not aware of the frames we are using. When we hear a word or observe a scene, we understand it by linking it to our existing knowledge or frame of reference. This helps us to filter all the information that we come across. For example, if we drive past a primary school at three in the afternoon and see a crowd of adults standing outside, we read the situation as parents waiting for the end of school and may look out for children in the road. A person from another culture may have a different frame, or filters, so they do not read the situation in the same way. In this example, they might ask what's going on and may even want reassurance that this crowd of adults is normal.

The sociologist Erving Goffman claimed that we have two types of frames:

- Natural frames that help us navigate the physical world, such as an ability to read the sky for weather or to know that something looks heavy. These neutral frames are neither good nor bad.
- Social frames that we learn through our interactions with other people as we are nurtured, flattered, threatened, and so forth. Our successes and failures will help us develop this sort of frame. These are not neutral.

According to cognitive scientist Jerome Feldman, our basic or primary mental frames are actually physical – they become fixed in the neural circuitry of the brain. In this way, our natural and social frames become integrated so we can understand phrases such as hot topic, which links an understanding of temperature in one part of the brain with references to conversation or media in another. Understanding all this is important because the frames that we live by will determine the way we view any situation, and will ultimately affect our behaviour.

One frame or metaphor we often encounter is thinking of the world as a complex machine, and this has become commonplace in industrialised societies that see machines every day. Using this frame, we might feel that the world's problems are technical and can be fixed with yet more technology. So, if burning fossil fuels has created global warming, we might aim to fix this by manufacturing an atmospheric shield using aerosols or by erecting vast mirrors in space to deflect the sun's rays. On the other hand, if we think of the Earth as a complex, self-organising system, like Gaia, we may wish to be more cautious in our management of the planet, perhaps addressing the original cause (burning fossil fuels) before attempting technical fixes.

It seems we are not the rational thinkers that we assume ourselves to be, as we do not simply respond to facts, but rather we think in terms of our frames. George Lakoff has written extensively about frames and has highlighted how businesses, politicians and the media use this understanding to powerful effect.[1] By framing situations in a particular manner, they can persuade us to share their interpretations. Communications expert Jim Kuypers suggests four ways that frames work; they are listed here along with an example taken from England's Neighbourhood Planning and Infrastructure Bill that was being discussed in parliament at the time of writing.

Notice how a new planning law is being framed in terms of an economic problem. In fact, the debate around this bill included reference to how the UK was no longer home to the world's largest port or airport (that's now in Dubai). This was seen as a problem despite the social, environmental and human health problems associated with large airports. In their book Framespotting, Laurence and Alison Matthews highlight numerous examples of how frames are used to influence public opinion.[2] One example they use is the phrase tax relief. This suggests that tax is a burden, something we need relief from rather than being thought of

Table 34.1 Examples of how frames are used

Purpose of frame	An example
Define problems	The economy is not growing fast enough to bring us prosperity.
Diagnose causes	Commercial development is hampered by planning restrictions that dictate where it can and cannot take place.
Make moral judgements	Local authority planners act badly when they put red tape before the needs of a vibrant (that is, successful) economy.
Suggest remedies	Reduce or remove planning restrictions so that development is free to take place in formerly restricted areas.

Source: Jim Kuypers[3]

as a membership subscription to a society that provides essential public services. They also discuss how economists are obsessed with economic growth as if the world is somehow undersize and infinite.

It seems that this framing game is often won by those who would exploit the planet for profit rather than by those who wish to manage it sustainably. George Lakoff claims that in the US, the Republicans are the ones who really understand framing. They have a cognitive strategy that ensures that 80% of talking heads on the media are putting forward conservative views; that is, proposing frames that they would like the public to share on a wide range of issues. Democrats, by contrast, still have what Lakoff calls an eighteenth-century brain. In other words, they still believe in reasoning, that if you tell people what is true or just, then they will follow it. As we have seen, reasoning isn't like that; we are informed by our frames.[4]

Many of those who want to promote sustainable development don't like the idea of manipulation, and recoil at the thought of a cognitive strategy, seeing it as akin to brainwashing. But if this is what happens naturally, if it is how we learn, then surely more of us need to get smart about framing, which is only dangerous when used for fraudulent reasons. Using it for the greater good cannot be bad unless we truly doubt that our interpretation is an honest one. Using framing to communicate what we understand to be the truth is perhaps the most honest thing we can do.

Notes

1 George Lakoff is retired Distinguished Professor of Cognitive Science and Linguistics at the University of California at Berkeley. He is Director of the Center for the Neural Mind & Society (cnms.berkeley.edu). Web link: georgelakoff.com.

2 Matthews, L. & Matthews, A. (2014). *Framespotting*. Alresford, UK: Iff Publishing.

3 Kuypers, J.A. (2009). *Rhetorical Criticism: Perspectives in Action*. Lexington, VA: Lexington Press.

4 In a February 2017 blog, Lakoff shows how to respond to criticism by reframing the argument in terms that are positive to your case. Web link: ow.ly/taTb309cGOX.

35

RESILIENCE

It is hard to open a book about sustainability or the environment without coming across the idea of resilience. WWF's 2016 *Living Planet Report* was no exception to this rule.[1] It offered this definition: "the ability of a . . . system to absorb and recover from shocks and disturbances, maintain functionality and service by adapting to chronic stressors, and transform when necessary."

Another view, this time from the Stockholm Resilience Centre, a group at the forefront of the development of ideas about resilience, is that it's "the capacity of a system . . . to deal with change and continue to develop."[2,3] Both definitions are about ecology, about the ability of the natural world to "continue functioning amid and recover from a disturbance," as the *Encyclopaedia Britannica* puts it.[4] The Stockholm Centre also says that resilience is about how humans and nature can use shocks and disturbances such as climate change to spur innovative social thinking.

This reminds us that the idea of resilience is used in everyday language to represent, more generally, the ability to withstand and recover from shocks and to keep working, either changing or not, as necessary. In this sense, resilience is not a quality you either have or don't have; it's one that you can have more or less of, and something you can develop or lose over time. It's unsurprising, then, that another everyday use is associated with psychology and our mental ability to cope what life throws at us.

Some of the most significant work on resilience was carried out by Canadian ecologist C.S. Holling, who distinguished between two types of resilience within natural systems.[5] The first, engineering resilience, concerns a return to equilibrium following shock. The second, ecological resilience, concerns adaptation to changed circumstances. The main difference between these two ideas is that where everything is working well, high engineering resilience results in minimal, if any, change, but high ecological resilience may result in change in one form or another because of the adaptation. A somewhat trivial, everyday example of a system with high engineering resilience would be an apple in the bottom of a deep fruit bowl, where tilting the bowl and then letting it go results in the apple returning to rest at the bottom of the bowl. Generally speaking, the faster such returns happen, the better. A good (and much less trivial) example of high ecological resilience would be an ecosystem that:

- contained a wide range of organisms;
- provided for many food chains and webs;
- had components that were dependent on each other; and
- contained species that could compensate for the loss of others.

High engineering resilience is a routine characteristic of the built environment – for example, bridges, earthquake-resistant walls and buildings, wind-resistant towers, and drainage and flood prevention structures – and part of so-called disaster risk reduction strategies. All such structures are built to a certain level of resilience within which all will likely be well because the system will cope with what is thrown at it. Beyond that, however, the system might well fail, perhaps catastrophically. In a general sense, high engineering resilience is always found in systems with high ecological resilience.

All natural systems have ecological resilience processes built into them. All mammals have it as an integral feature of how their body systems work. For example, our internal body temperature is maintained within a few degrees of fatal upper and lower limits, and our body systems adapt to changed circumstances to keep us within these limits. Surprisingly perhaps, machines can also have high ecological resilience; modern car engine management systems make second-by-second adjustments to fuel and air mixtures to keep performance at optimal levels, and fighter jets can fly at high speed and low altitude constantly adjusting to changing circumstances.

The idea of resilience has been used as metaphor or tool to help us think about a wide range of problems. According to the Stockholm Resilience Centre, there needs to be what they term resilience thinking (by humans) if the resilience of natural system is to be enhanced. They argue that such thinking is concerned with:

- generating increased knowledge about how we can strengthen our capacity to deal with the stresses caused by environmental change; and
- finding ways to deal with unexpected events and crises, and finding sustainable ways for humans to live within the Earth's boundaries.

The Stockholm Centre says that resilience thinking has three components:

1   There are complex interdependencies between people and ecosystems because our economies and societies depend on those ecosystems to provide fresh air, clean water, good food and a huge range of other goods and services.
2   It is human development over the past 200 years that has brought the planet dangerously close to the point where abrupt environmental change cannot be excluded, and has led to the idea that we have entered a new geological era called the Anthropocene, where humanity is influencing every aspect of the Earth.
3   The paradox that our intelligence and capability, although it has led to our environmental predicament, can also be used to get us out of it.

In short, resilience thinking is necessary if we are to find innovative ways to reconnect with the biosphere and stay within planetary boundaries. It is certainly the case that long-term Sustainable Development Goals are rendered meaningless when human life, environments and infrastructure are devastated by natural disasters, and so it's critical that communities are resilient; that is, they have adaptive capacity, defined as the ability to better prepare for, and learn from, catastrophe. Education is seen as an important strategy in building such resilience, and the Hyogo Framework for Action (2005–2015) calls on the international community to integrate disaster risk reduction (DRR) within education policies.[6] The purpose of this is so that communities examine their vulnerabilities to hazards, and rebuild in ways that enhance safety, especially for their most vulnerable members.

If resilience is a good thing in general, it needs to be said that high resilience is not always a desirable feature of a system. For example, a lake that contains too much nitrate, perhaps from agricultural overuse, can result in very low oxygen levels, which means that fish species cannot live there, while other species can. The resulting ecosystem might be resilient, but it's not the sort of ecosystem that either humans – or fish – want. In other words, our values always come into play when we think about the sort of system and circumstances that we want to see or preserve.

Notes

1 WWF (2016). *Living Planet Report*. Web link: ow.ly/THlX3o8LVLN.

2 Stockholm Resilience Centre. Web link: stockholmresilience.org.

3 The Stockholm Resilience Centre's video explaining resilience can be seen online. Web link: ow.ly/SzV63o8LVCL.

4 *Encyclopaedia Britannica* article on resilience. Web link: britannica.com/science/ecological-resilience.

5 Holling, C.S. (2010). The resilience of terrestrial ecosystems. In: L.H. Gunderson, C.R. Allen & C.S. Holling (eds), *Foundations of Ecological Resilience*. Washington, DC: Island Press.

6 Hyogo Framework for Action (HFA). Web link: unisdr.org.

36

COMPLEXITY

We have included a brief look at complexity here because it can help us get to grips with the frankly weird nature of some of the interrelationships that characterise our world. These include both ecological systems and globalisation. It also has huge technological implications that may yet help us resolve some of our pressing problems.

There is no single theory about complexity; it crops up in many fields, from archaeology to psychology. It developed from catastrophe theory, which suggests that a butterfly flapping its wings in Argentina can cause a hurricane in the Caribbean, and from chaos theory. It's certain that as we explore more complex systems, we will better understand them, but it is unlikely that we will ever be able to *predict* their outcomes. This is disconcerting because, as we discuss in our systems chapter, we have been raised with a linear framing of ideas and ways of thinking that assumes straightforward cause and effect relations. But such complex relationships are often unstable right down to the quantum level, where electrons and photons can exist as a combination of states that correspond to different outcomes. Such a quantum system remains like this until it interacts with the external world, for example when it is observed. Only then does it adopt one of the possible states. Einstein said this was "spooky."[1]

While systems thinking highlights the interconnected nature of our world, complexity takes this a step further; phenomena react to and with their environments. In the 1970s, the idea of biologist Richard Lewontin that species do not simply adapt to environments, but construct them, led us to rethink key aspects of evolution.

With different components acting in multiple directions, complex systems can offer an almost infinite range of possibilities, so it's not surprising that simple systems can quickly develop higher levels of complexity. This tendency towards increased complexity demonstrates the core principle of self-organisation. Complex systems often operate without a decision-maker, conductor or driver. Examples include shoals of fish and murmurations of starlings that soar and swoop as if a single organism. In fact, the individuals are following a limited set of instructions, such as keeping a certain distance away from neighbours.[2]

Another dimension of self-organising systems is that they are autopoietic— literally, self-creating. Feedback causes change, which in turn creates feedback and more change. In the early 1990s, the Chilean scientists Humberto Maturana and Francisco Varela identified autopoiesis as a defining feature of life itself.[3] This process was demonstrated in the case of human biology in 2016, when laboratory studies of 13-day-old embryos showed them taking shape independently of any maternal tissue, highlighting their remarkable and unanticipated self-organising properties.

At the planetary scale, autopoiesis applies to James Lovelock's concept of Gaia, where life on Earth creates, maintains and recreates the very conditions necessary for life to continue. This brings us to another dimension of complexity: the way that relationships repeat themselves regardless of scale. This gives rise to fractals in nature such as the recurring patterns in different scales of drainage systems or the fronds of ferns. These also occur in human infrastructure such as in communication networks and financial systems.

Closely linked to self-organisation is the central idea of emergence, which happens when elements self-organise to create completely new phenomena. Life, mind and fire are often cited as familiar cases of emergent properties because they cannot be predicted from a study of their component parts. This is because when those parts are put together, a genuinely new phenomenon appears.

A stock market is an example of a complex system where people buy and sell individually with no central directing force. The actions of individual traders affect the movements of the market, which in turn affects the actions of the traders. We might expect this to arrive at a steady state, but small changes can have large and sudden effects, hence bubbles and crashes occur that microeconomics cannot explain. Booms and recessions, it seems, are emergent properties of the capitalist system.

This leads to another feature of long-lasting complex systems: they are always poised at some point between order and chaos, and standing still is not a survival strategy. A system that is too ordered and rigid becomes brittle and is as likely to collapse as it is to respond or evolve. We see this when we interfere with natural systems. For example, wildfires in nature are a self-organising mechanism that avoids increased order that might otherwise create this brittleness. If we prevent natural fires from occurring, then the environment may accumulate so much combustible material that when a fire does happen, its ferocity will far exceed the strategies employed by plants and animals to survive such events. Similarly, globalised economic systems exist between order and chaos as the ebb and flow of trade and communication encourages constant innovation. Attempts to control the system, for example through trade barriers, can give the appearance of stability, but ultimately also creates brittleness and possible collapse, as seen in the Great Depression of 1920s USA.

What does this mean for sustainable development and our future? This flexible way of understanding our world can help us to rethink the limits and overlaps of economic, political and social systems. To date, our limited view, focused on nation states, has failed to grasp the multidimensional and interconnected nature of the issues we face. Only if politics and economics can adopt a more complex, non-linear view will we be able to look forward to increasingly nuanced and effective long-term strategies. At the individual level, the idea of autopoiesis suggests that everything we do contributes in a small way to remodelling the system of which we are a part. The system is not something that is done to us; we are both products and producers, so the world we see is not as inevitable as we might think.[4]

At the start of this chapter, we referred to technological implications; the most significant of these will probably be the development of quantum computing. Today, our computers operate using algorithms made up of strings of binary bits: 0s or 1s. A quantum computer is powered by particles (qubits) that can be 1, 0 or neither. This unleashes vastly increased computing power. Although there are enormous technical hurdles to overcome, not least the difficulty of extracting answers from an unstable complex system, the first two-qubit quantum computers are in operation. Assuming there are solutions to our most intractable problems, these machines will soon be on the case.

Notes

1 Erwin Schrödinger illustrated the idea of a combination of states and wave function collapse with a thought experiment of a cat sealed in a box with a capsule of cyanide which might be about to be smashed (or not) because of radioactive decay. As a result, the cat is both dead and alive – until we open the box to find out which it is. Wikipedia has a brief introduction to this idea. Web link: ow.ly/hzN5308ZclP.

2 Shahbazi, N.M. et al. (2016). Self-organisation of the human embryo in the absence of maternal tissues. *Nature Cell Biology*, 4(533), 251–254. Web link: ow.ly/d42j308ZAu8.

3 Maturana, H. & Varela, F. (1994). *The Tree of Knowledge: The Biological Roots of Human Understanding*. Boulder, CO: Shambhala.

4 Pueyo, S. (2014). Ecological econophysics for degrowth. *Sustainability*, 6, 3431–3483. Web link: ow.ly/6Z1C308ZAhk.

37

GLOBALISATION

A straightforward dictionary definition of globalisation is making something worldwide in scope or application. Whether viewed broadly in relation to business, or more narrowly in terms of finance and investment, globalisation has many implications for the relationship between people and nature, and for sustainable development.

In the broad sense, globalisation means that we can now buy the same brands of terrible beer and fizzy drink in almost every country in the world. We can see this form of globalisation almost everywhere. It has come about because marketplaces that were once local are now increasingly global. Not only are goods that were once associated with a particular country or culture now available in many countries; often they are made in those different countries. This is striking in relation to food, particularly fast food. But it is also marked when it comes to consumer and other goods. For example, Toyota sells its vehicles in 170 countries, and produces them in 28, from Argentina to Vietnam.

This happens because it makes economic sense to Toyota and its investors. It's not just that it's easier to do this than ship the cars by sea from Japan. Manufacturing abroad means that far more cars can be produced, and are free of import duties and tariffs in those countries. We see the same phenomenon in relation to aircraft and aero engines, where component parts are sourced from specialist companies across the planet. The same can be said of manufacturers of smartphones, games consoles, TVs, etc., and increasingly of services such as banking and insurance. Such companies tend to be referred to as multinationals, many of which not only trade internationally, but have international investors and owners too.

This aspect of globalisation, which might be characterised as trading and investing across borders, is nothing new. This buying and selling across great distances began with the domestication of the camel around 1000 CE and led to the creation of the Silk Road across Central Asia linking China and the Mediterranean. The great ocean voyages of the late Middle Ages followed, opening the world up to trade and European colonisation, and the invention of the steam engine in the early 1700s added a huge techno-logical boost to industry and trade. These are all examples of what Richard Baldwin terms "unbundling," in which consumption was separated from production, and was the first phase of globalisation. The second phase, he says, has just happened, now that information, money and ideas can be transmitted across the globe at high speed and low cost.[1]

But globalisation embodies more than just international trade and investment, which might be argued to be positive, provided you see trade as more than a zero-sum game. It is very clear that globalisation is a process where there are losers as well as winners. Overseas ownership of a coun-try's assets implies that their owners might not always have the interests of the country and its population as their first priority, and the international nature of business now is such that investment can be withdrawn from a country as quickly as it can be placed there. It follows that jobs can some-times disappear as fast as they are created as capital is shifted to where return on investment is better. Because of this, lower costs in one country can lead to entire industries vanishing in another. This was the case with shipbuilding in the UK, though naval vessels are still built there. It's also true of basic steel-making in the US, although high-value specialist steel production remains because of the prices it can command.

A workforce on the receiving end of all this tends to grumble – assuming it remains a workforce at all – even though liberal economists might say that it's rather inevitable given how easily capital now travels round the world. Governments whose people are affected by such globalised events are usu-ally faced with the choice of subsidising the job or the worker. Doing the first means funding the industry to try to keep the existing jobs as long as possible; doing the second means supporting workers and possibly their communities in the short term, helping them retrain and easing investment for new industries and jobs. Attitudes to these choices depend on prevailing political cultures, and a huge contrast can be found between countries. UK governments have tended since the 1980s to recognise economic inevitability when they see it, even when whole communities are involved – in the loss of steel-making, for example. By contrast, French governments, which sometimes seem to think that globalisation threatens what it is to be French, tend to resist foreign ownership of business, even though French multinationals are some of the most adept at exploiting globalisation for the benefit of their shareholders. In 2005, Danone, a maker of yoghurt, was declared a strategic national asset by the French government in order to fend off the interest of PepsiCo. Foreign ownership of business, especially by fleet-footed multinational ones, tends also to create tensions when it comes to profits and taxes, and there is a popular feeling in many countries that too many of such companies move money around in order to pay as little tax as possible where they actually trade.

If international trade is nearly as old as human history, that cannot be said about how we experience globalisation today. This is driven by technology that has shrunk distance and time, and by the ability to create money out of nothing, and transfer it at (almost) the speed of light to anywhere on Earth. This enables a call centre to be thousands of miles away from those calling, products to be designed in one country, made in another, and sold everywhere, and data to be stored on servers half the globe away.

An important dimension of this process is cultural globalisation. This is the transmission of ideas, values and cultural artefacts such as popular music, the films of Hollywood and Bollywood, and news outlets with, of course, most of the content distributed via the internet. This is often seen as a homogenising force, if not outright cultural imperialism, when carried out by well-resourced Western manufacturers of content. It's not so simple, of course. All of us have a cultural identity, and this is coloured, if not shaped, by globalisation, which we in turn contribute to through our multiple interactions. And if ideas and values such as those associated with sustainable development are going to spread, it will be through this very process.

Globalisation also encourages the harmonisation of laws and regulations across countries to reduce barriers to trade, and international trade agreements these days are as much about this, as they are about removing tariffs.

But this comes with a social price, as such cooperation is both economic and political. Economist Dani Rodick argues that, because of this, countries now cannot be globally integrated, completely sovereign, and democratic.[2] Two of these are possible, he says, but not all three.

As Newton once explained, for every action, there is a reaction, and it's no surprise that globalisation has spawned anti-globalisation movements that oppose what they see as a surrender of power to unaccountable multinationals. Such groups think that international trade agreements and deregulated financial markets give too much power to companies, which they say try to maximise profit at the expense of safety, conditions of service, wages, and the environment. Activists call for forms of global business that don't undermine democratic representation, human rights, fair trade and sustainable development. This movement is a broad church, ranging from neo-Luddites who want life to be simpler, some of whom are prepared to use violence to achieve this, to those who think that it's global trade itself that's the problem, across to those motivated more by the global pursuit of fairness and social justice. Of course, not everyone agrees that globalisation is bad, and many argue that spreading markets and trade into the developing world is the best way to beat poverty, providing there is sound regulation.

One unwanted facet of globalisation is that the travel and trade it involves makes the spread of disease easier, and global pandemics more likely. As Ali S. Kahn and William Patrick note in a recent book, globalisation is spreading infection round the world like never before.[3] These are unlikely to be exotic diseases, but more mundane ones such as flu, including the spread of antibiotic-resistant bacteria. It was appropriate in every sense, therefore, for Bill Clinton to say that globalisation was the economic equivalent of a force of nature, like wind and water.

Notes

1 Baldwin, R. (2016). *The Great Convergence: Information, Technology and the New Globalisation*. Cambridge, MA: Belknap.

2 Rodrik, D. (1997). *Has Globalisation Gone Too Far?* Washington, DC: Institute for International Economics.

3 Kahn, A.S. with Patrick, W. (2016). *The Next Pandemic: On the Front Line Against Humankind's Gravest Dangers*. New York: Public Affairs.

38

NEOLIBERALISM

This is a catch-all term that is often used pejoratively to describe a range of economic policies. It represents a transfer of power from democratic institutions to the corporate sector, and from labour to capital, where globalised business is increasingly free of national constraints and accountability, and has the legal means to take on governments that get in its way, and where the individual is emphasised rather than their role in society.

The payment for higher education in England illustrates the issues. English universities charge fees to students who then graduate with debts. The student is seen as the individual who is the one who benefits most from the experience. The alternative communitarian view, such as prevailed in the 1960s and 1970s, sees higher education as a universal benefit that adds to the collective intelligence of society, and as something to which we should all contribute through our taxes. Some communitarians say that the taxpayer should simply fund higher education as an investment in society. However, others argue that if students go on to earn high salaries, they should pay more through a graduate tax than those who didn't go to university. In the end, of course, there might not be much difference in money terms between paying a graduate tax and paying off tuition fees. To neoliberal purists, however, such an approach with its taxpayer subsidies is just the sort of state interference between the individual and the market that they deplore.

Critics of neoliberal thinking say that it leads to a low-trust, dog-eat-dog world. They claim that it means that governments abdicate their responsibilities, reducing everything to a series of individual decisions. The result, they argue, is that an advantaged elite accumulates wealth, while the poorer and less able face a precarious existence of job insecurity and low wages.

To see where this strain of neoliberal thinking originated we first need to consider liberalism itself. This political idea grew out of seventeenth-century religious disputes and the Enlightenment. This was a period from the seventeenth to nineteenth centuries, when scientific discoveries ushered in new technologies and ways of thinking that undermined notions of divine and/or autocratic rule by priests and monarchs. The idea that people had inalienable rights was promoted by thinkers such as John Locke, and later by Thomas Paine, who called for society's liberalisation from inherited government. Such principles underpinned the constitution of the USA. Paine's contemporary, Adam Smith,[1] author of The Wealth of Nations, argued that commerce should be allowed to take place unfettered by the interference of governments in order to maximise the wellbeing of everyone and as a way of ending wars. Here, the market was an efficient means to an end: that of human wellbeing. This is a similar argument as that put forward by Daly and Meadows in their ideas around the four capitals (see Chapter 31). With all this emphasis on markets, there was also the idea that individuals should be free to live their lives in ways that they value.

However, as the Industrial Revolution gained momentum, human wellbeing and happiness took a back seat as laissez-faire (letting things take their own course) approaches dominated a competitive dash for wealth. The result was squalid living conditions, child labour in factories, and adult workers being paid starvation wages while trapped in debt. This drew widespread criticism, most significantly from Karl Marx, who attacked the way that laissez-faire led to the destitution of workers. There was also opposition from within business from social reformers such as Lever, Rowntree, Dale, Reckitt and Fry, who treated their workforces comparatively well, providing housing, education and modest social welfare. By the twentieth century, organised labour had helped to improve working conditions, and the excesses of laissez-faire were checked through legislation. The economic depressions of the 1930s further illustrated the shortcomings of market-led approaches as governments felt obliged to intervene on behalf of the poor. It was at this time, however, that some expressed concern about the implications of over-dominant government. A key figure was Friedrich Hayek, who published *The Road to Serfdom* in 1944.[2] Hayek argued that central government planning inevitably leads to totalitarianism, and that government's only legitimate role should be upholding the rule of law, allowing individuals to go about their business freely.

Such views were not popular. Instead, post-Second World War policies sought consensus between the state, capital and labour. Governments followed the economic ideas of John Maynard Keynes that sought to avoid the recessions of the business cycle by borrowing money to support activities such as infrastructure projects. Because of this approach, government borrowing was high, and when oil prices quadrupled in the 1970s, economic difficulties soon followed. New governments led by Margaret Thatcher (UK, 1979) and Ronald Regan (US, 1980) took the views of Hayek and others and made them mainstream. The old post-war consensus was replaced with a new order in which capital (financial and corporate power) became dominant. The term neoliberal was coined in 1980s Latin America, where deeply indebted governments were forced to introduce a range of liberal policies by the International Monetary Fund (IMF).[3] Similar measures were prescribed across Africa in the form of structural adjustment programmes in the 1990s.

In 1989, the English economist John Williamson coined the term Washington Consensus to describe a list of 10 policy prescriptions that would be agreed upon by institutions such as the IMF and World Bank, based in Washington, DC.[4] This is a summary:

1 Budget discipline, avoiding large financial deficits.
2 Redirection of public spending from indiscriminate subsidies towards broad-based services such as education, healthcare and infrastructure investment.
3 Broadening the tax base with moderate tax rates.
4 Interest rates that are market-determined.
5 Competitive currency exchange rates.
6 Trade liberalisation with low tariffs.
7 Liberalisation of inward foreign investment.

8   Privatisation of state enterprises.
9   Abolition of regulations that impede market entry or competition, except on safety, environmental and consumer protection grounds, or for the oversight of financial institutions.
10 Legal security for property rights.

This has been the trickiest chapter to write because we differ in our view of all this. What you think of the Washington Consensus will likely depend on your politics, although some points, such as the security of property rights, really ought to be uncontentious these days – except where indigenous people continue to be dispossessed by those who have acquired some legal title over their territory. You might see no merit in the consensus at all, only problems. You might quite like some of it, up to a point, or quite like it all, but only if not carried out to extremes. We do agree, however, that in relation to sustainability, a full neoliberal approach would embody a range of risks, including that:

• pollution and waste (externalities) would be left for others to pay for;
• removing too many regulations would weaken social and environmental protection;
• global agreements would be hard where forest or water management has been privatised;
• short-term thinking can result in the maximisation of profits with no regard for anything else; and
• when corporations have individual rights, they can behave in ways that most socialised human beings wouldn't.

Most dangerously, some argue, the ideal of an endlessly competing work-force denies the relational nature of life and our social connections. Pankaj Mishra puts it like this in his book *Age of Anger: A History of the Present*:

In advanced democracies, a managerial form of politics and neo-liberal economics has torn up the social contract. In the regime of privatisation, commodification, deregulation and militarisation, it is barely possible to speak without inviting sarcasm about those qualities that distinguish humans – trust, co-operation, community, dialogue and solidarity.[5]

If we are to learn how to deal with our sustainability predicament, we will need to collaborate in ways that a blinkered faith in markets does not allow. That said, the real problem might not be the market, but the blinkered faith.

Notes

1 Smith, A. (1776). *An Inquiry into the Nature and Causes of the Wealth of Nations*. London: W. Strahan and T. Cadell.
2 Hayek, F. (1944). *The Road to Serfdom*. London: Routledge.
3 IMF (2003). Beyond the Washington Consensus. Web link: ow.ly/fYZl308LWMn.
4 Wikipedia has an introduction to the Washington Consensus. Web link: en.wikipedia.org/wiki/Washington_Consensus.
5 Mishra, P. (2017). *Age of Anger: A History of the Present*. London: Alan Lane.

PART III

STRATEGIES

39

REDUCING GLOBAL INEQUALITIES

In 2000, the United Nations Millennium Summit set the world eight ambitious goals to be achieved by 2015, and all member states, together with a range of global organisations, committed themselves to help achieve what became known as the Millennium Development Goals (MDGs). Table 39.1 on the next page shows the goals and some of the progress made.[1]

Despite the sort of good news statistics shown above, progress was really very uneven, and in some countries hardly any was made. What was probably the most important target – to halve the proportion of people living on an income of under $1.25 a day – was achieved well before 2015, with over 700 million people being lifted out of extreme poverty. But this is a global figure, and there are huge variations between regions and countries. The best progress was in China, where the proportion fell from 60% in 1990 to 12% in 2010. By contrast, in sub-Saharan Africa, it only fell from 56% to 48%.

Ban Ki-moon, the previous UN secretary general, acknowledged that while there had been some positive results, it still left too many people behind. No matter how much progress has been made in many areas, it can

Table 39.1 The Millennium Development Goals and some of the progress made

Goal	Progress made
1 Eradicate extreme poverty and hunger	In 1990, 47% of the population in the developing world lived on less than $1.25 a day. In 2015, it was 14%. Globally, the number of people living in extreme poverty fell from 1,900 million in 1990 to 836 million in 2015.
2 Achieve universal primary education	Primary school enrolment in the developing world reached 91% in 2015, up from 83% in 2000. Globally, the number of primary age children not in school fell from about 100 million in 2000 to 57 million in 2015. The literacy rate among young people aged 15 to 24 has increased globally from 83% (1990) to 91% (2015).
3 Promote gender equality and empower women	Many more girls are now in school compared to 15 years ago. In southern Asia, for example, only 74 girls were enrolled in primary school for every 100 boys in 1990. Today, it is 103 girls.
4 Reduce child mortality	Globally, the under-five mortality rate declined from 90 (in 1990) to 43 deaths per 1,000 live births in 2015. Despite continuing population growth, the number of deaths of children under five has declined from 12.7 million in 1990 to about 6 million in 2015.
5 Improve maternal health	Since 1990, the maternal mortality ratio has declined by 45% worldwide; now, there are 210 deaths per 100,000 live births. Globally, over 70% of births were assisted by skilled health personnel in 2014 – up from 59% in 1990.
6 Combat HIV/ AIDS, malaria and other diseases	New HIV infections fell by about 40% between 2000 and 2013, from about 3.5 million cases to 2.1 million. By 2014, 13.6 million people living with HIV were receiving antiretroviral therapy, up from 800,000 in 2003. The global malaria mortality rate has fallen by 58%, and, between 2000 and 2013, tuberculosis prevention, diagnosis and treatment saved an estimated 37 million lives.
7 Ensure environmental sustainability	Some 1.9 billion people have gained access to piped drinking water since 1990. The total is now 4.2 billion. Worldwide, 2.1 billion people have gained access to improved sanitation. The proportion of people practising open defecation has fallen almost by half since 1990.
8 Develop a global partnership for development	By 2015, 95% of the world's population were covered by a mobile cellular signal.

Source: UN

be easy to forget what still needs to be done. For example, saying that the average proportion of women in national parliaments has nearly doubled over the past 20 years sounds like good progress, whereas saying that, on average, only one in five members are women, tells a different side of the same story.

China and India's reductions of extreme poverty were brought about the sort of rapid economic growth that Africa's poorest countries can only, for now at any rate, dream of, but there is a general trend here which shows that where money can be spent on straightforward technologies, such as water purification, it can be effective. Similarly, money spent on public health programmes clearly contributed to the drop in deaths from diseases. That said, in 2014, there were still 115,000 deaths from measles, and almost as many from cholera. An estimated 1 million children became ill with TB in 2014, and 140,000 died of it. Much worse, according to the World Health Organisation, even though an estimated 6.2 million malaria deaths have been prevented since 2001, there were 214 million cases in 2015 and 438,000 deaths. Given all this, it was no surprise that the UN decided to build on the limited success of the Millennium Development Goals, and develop more ambitious targets for 2030. These are the Sustainable Development Goals (SDGs), often referred to as the Global Goals.

There were just eight Millennium Development Goals (MDGs), tightly focused on cutting extreme poverty and improving healthcare and education, and all were reasonably clearly defined. By contrast, there are 17 SDGs with 169 targets, covering world peace, the environment, gender equality and more. Many are thought to be impossible to measure effectively, but at their launch in 2015, Ban Ki-moon said: "Reflecting on the MDGs and looking ahead to the next 15 years, there is no question that we can deliver on our shared responsibility to end poverty, leave no one behind and create a world of dignity for all."[2]

The Sustainable Development Goals include the familiar and sharply phrased "End poverty in all its forms everywhere," which would be possible to measure, but probably impossible to achieve; and the new but less succinctly put "Build resilient infrastructure, promote inclusive and sustainable industrialization, and foster innovation," which might be possible to achieve, but is probably impossible to measure.

Unfortunately, most of the goals are phrased like the second example here. Indeed, they read like aspirations rather than goals, although some of the targets associated with the goals provide clear, valuable and measurable outcomes. A particular criticism of the SDGs is that they are a mix of the cost-effective and the costly, and the Copenhagen Consensus Centre, an international think tank, would like to see a focus on a smaller number of the 169 targets, which would have a large impact on the problems.

Michael Anderson, CEO of the Children's Investment Fund Foundation. was quoted in *The Economist* as saying:

> The MDGs were meant to create a social safety net; the SDGs [are meant] to be fit for an age in which the standard of living in a big chunk of the developing world is creeping towards the levels of rich countries.[3]

Although the SDGs are less clear than they might have been, they do a much better job of covering the interrelated economic, social and environmental issues the world faces. In that sense, they are more useful than the MDGs were, and they are a useful way of focusing the whole world's attention on what should be its priorities.

Notes

1 UN (2015). Millennium Development Goal (MDG) Report. Web link: ow.ly/5FVA308S1TG.
2 Foreword to the UN Millennium Development Goals Report (2015, p. 2). Web link: ow.ly/BcME30g3FTa.
3 *The Economist* (2015). The good, the bad and the hideous. 26 March. Web link: ow.ly/GRCr308S2jz.

40

THE COPENHAGEN
CONSENSUS

The United Nations' Sustainable Development Goals (SDGs) were agreed
in 2015.[1] There are 17 goals and 169 separate targets. There is, according to *The Economist*, something there for everyone. This is because the
UN involved not just countries and UN agencies in the goal development
process, but a huge range of non-governmental organisations (NGOs) as
well. These were keen to contribute their ideas, and equally keen to make
sure that they were adopted. As a result, the range of goals and targets
is both extensive and occasionally baffling. For example, Target 7 of
Goal 4 (ensure inclusive and quality education for all and promote lifelong
learning) is, by 2030, to:

> Ensure all learners acquire knowledge and skills needed to promote sus-
> tainable development, including among others through education for
> sustainable development and sustainable lifestyles, human rights, gen-
> der equality, promotion of a culture of peace and non-violence, global
> citizenship, and appreciation of cultural diversity and of culture's contri-
> bution to sustainable development.

As *The Economist* said, "Try measuring that."[2] This is not an isolated example. Others include:

Goal 1 Ending Poverty – Target 6:

Ensure significant mobilisation of resources from a variety of sources, including through enhanced development cooperation, in order to provide adequate and predictable means for developing countries, in particular least developed countries, to implement programmes and policies to end poverty in all its dimensions.

Goal 12 Ensure sustainable consumption and production patterns – Target 11:

Rationalise inefficient fossil-fuel subsidies that encourage wasteful consumption by removing market distortions, in accordance with national circumstances, including by restructuring taxation and phasing out those harmful subsidies, where they exist, to reflect their environmental impacts, taking fully into account the specific needs and conditions of developing countries and minimising the possible adverse impacts on their development in a manner that protects the poor and the affected communities.

The problem with the sustainable development goals and targets, some argue, is not their lack of worth or importance, but that there are so many of them. If anything is to be achieved, they say, more focused attention is needed. This is something that the Copenhagen Consensus, an international think tank, has been looking at.[3] The Consensus says that establishing priorities for action is crucial if increases in global welfare, as embodied in the goals and targets, are to be achieved. Understanding that resources are scarce, the Copenhagen Consensus uses the theory of welfare economics and cost–benefit analysis to focus on what can be achieved most effectively.

It asked economists and researchers to look at the goals to see what might be the most cost-effective ways to achieve the targets. It found that 18 of the 169 targets would pay back $15 or more for every $1 invested. The 18 are set out in Table 40.1 on the next page.

The Copenhagen Consensus methodology is to ask experts to come up with possible solutions to problems. These are then evaluated and ranked by a panel of economists using cost–benefit analysis to focus on maximising the benefits that might be obtained. This approach, which some may see

Table 40.1 Examples of cost-effective ways of meeting the Sustainable Development Goals[4]

Proposal	Goal	Investment ($bn)	Benefits
Lower chronic child malnutrition by 40%	2	11	Prevent malnutrition in 68 million children a year
Halve malaria infection	3	0.6	Save 440,000 lives a year; prevent 100 million cases of malaria
Reduce tuberculosis deaths by 90%	3	8	Save up to 1.3 million lives a year
Avoid HIV infection through circumcision	3	0.035	Avert 1.1 million HIV infections by 2030
Cut early death from chronic diseases by 1/3	3	9	Save 5 million lives a year
Reduce newborn mortality by 70%	3	14	Prevent 2 million newborn deaths a year
Increase immunisation to reduce child deaths by 25%	3	1	Save 1 million child deaths a year
Make family planning available to everyone	3	3.6	Cut maternal deaths by 150,000
Eliminate violence against women and girls	5	?	Cut cases of domestic abuses by 305 million
Phase out fossil fuel subsidies	12	<37	Free up $550 billion of government spending a year
Cut indoor air pollution by 20%	3	11	Prevent 1.3 million deaths a year
Complete the Doha round of trade liberalisation	2	20	Reduce extreme poverty by 160 million; a 10% increase in global income
Improve gender equality in ownership, business and politics	1	?	?
Boost agricultural yield increase by 40%	2	2.5	$84 billion gain; 80 million fewer hungry people
Increase girls' education by two years	4	?	?
Universal primary education in sub-Saharan Africa	4	9	30 million more African children attend primary school
Triple preschool in sub-Saharan Africa	4	6	30 million more African children attend preschool
Halve coral reef loss	14	3	An additional 3 million hectares of coral reef

Source: Copenhagen Consensus

as far too rationalist and neoliberal, is justified as a counter to established practice in international development. Here, it's claimed public opinion and the media conspire to set priorities that are often unachievable, and which result in less progress than might be achieved at a higher cost.

You don't have to accept all the economic logic of the Copenhagen Consensus methodology to find some value in this sort of exercise. At the very least, it provides information to assist in the wise spending of scarce money. It also forces you to acknowledge that there are always issues to be faced. These are not whether we want 80 million fewer hungry people every day or 30 million more African children to attend primary school, a cut in maternal deaths by 150,000 or 1.3 million fewer deaths a year from tuberculosis. Most of us are likely to want all these and more. The issue is whether we're prepared to prioritise, and therefore compromise, now that there are 169 targets to be met.

Notes

1 UN (2015). The Sustainable Development Goals. Web link: ow.ly/uSKo308LYcq.
2 *The Economist* (2015). The good, the bad and the hideous. 26 March. Web link: ow.ly/GRCr308S2jz.
3 Lomborg, B. (2013). *How to Spend $75 Billion to Make the World a Better Place*. Copenhagen: Copenhagen Consensus Centre.
4 The use of "?" in the table indicates an absence of data.

41

FEEDING 10 BILLION

Thomas Robert Malthus was an important economist and is remembered (at least by economists) for a series of debates with David Ricardo in the 1820s about political economy. Non-economists tend to know about him because of a pamphlet he published in 1798. In this, he argued that population growth would outrun agriculture's ability to satisfy human hunger, and these arguments have never quite gone away. The publication had a fine title: *An Essay on the Principle of Population as it Affects the Future Improvement of Society, with Remarks on the Speculations of Mr. Godwin, M. Condorcet, and Other Writers*. This was republished several times up to 1826, and at heart was an attack on the beliefs of those of a utopian frame of mind (such as William Godwin) who felt that life could, and would, continue to improve for all of humanity.[1]

Malthus was convinced that the availability of good agricultural land was limited, and so would not be able to keep up with the number of people demanding to be fed. And so, despite a lot of factors that reduced birth rates, such as not marrying or doing so late in life, choosing not to have children, birth control (and homosexuality), and factors that increased the death rate, such as disease, war, disaster, drought and famine, Malthus thought the outlook gloomy, and advocated social reform to persuade families from having too many children.

Malthus's argument was a simple mathematical one. He said that human population, if unchecked, would increase in a geometrical way; that is, it would double and redouble in any given time period (1, 2, 4, 8, 16, . . .), say every 25 years. Whereas, Malthus argued, our ability to produce more food could only grow at an arithmetic rate in the same time period; that is (1, 2, 3, 4, 5, . . .). Because of this, he argued that agriculture could not support a growing population, and the result would be poverty, malnutrition, starvation and death on a large scale.

Although his ideas were influential, they have not (yet) proved to be correct. He was writing as the Industrial Revolution was developing, and it turned out that land supply was not the limiting factor he supposed it to be. Rather, improvements in agricultural practice, particularly through the use of machinery, plant breeding and biochemistry (fertilisers, pesticides), resulted in given acreages of land producing an increasing amount of food. A stark illustration of these developments is provided by World Bank figures on global grain yields, showing a rise from just over 1,400 kilograms per hectare in 1961 to almost 3,900 kilograms in 2014.[2]

For economists who follow Malthus's ideas today, natural resource scarcity remains a key idea, even though the human disaster that was foreseen has yet to come to pass. And if you do think about such matters from a scarcity perspective, you do tend to see, as Malthus did, that reducing population is the key. And many today, faced by a rapidly changing climate, remain worried that agriculture will fail at some point, or, perhaps more accurately, do not see how it can continue to cope with a growing world population that will soon be 10 billion or so.

It might seem odd to claim that agriculture has continued to deliver the goods when so many have in fact died in famines over time. The 30 million people (or thereabouts) who died in Mao's Great Leap Forward in the late 1950s did so largely because they did not have enough to eat. But that was the fault of politicians who deliberately destroyed agricultural production to create space, literally and metaphorically, for industrial development. Similarly, the Ethiopian famine (1983–1985) that gave rise to the Food Aid and Band Aid campaigns in 1984 was in large part caused by government policies leading to social and political breakdown, rather than an absolute lack of food in the region.

The Nobel Prize-winning economist Amartya Sen has said that famines tend not to occur in liberal democracies.[3] He did not mean that no one ever goes short of food, or is ever malnourished, as in many economically developed countries there always seem to be people at the margins of society who do get not enough food, even though it is plentiful. Rather, he was making a point about communications and responsiveness. In well-functioning liberal democracies, there will be a free press ever keen to dish the dirt on the government and willing to point to problems, and those governments will, generally speaking, want to be seen to do something about such problems in order to get re-elected. This is not a recipe for utopia, but it does embody political checks and balances that help to identify severe problems quickly. After his Great Leap Forward had failed, Mao was fond of saying that if only he'd known about the problems, he could (and would) have fixed them.

If you ask those who say that something must be done about the global population, a usual response is to say that education is the key to changing how people think about the size of their families. Others say that increased affluence will be a faster method, as being better off, and more secure, usually results in fewer children. And if increased affluence brings better healthcare, then more children survive into adulthood.[4]

It's quite unusual to find those saying that something must be done actually advocating practical steps, and China's discredited one-child policy is probably the best recent example of such an attempt. India has tried forced sterilisation in the past, but neither of these approaches fits well with the norms of a liberal democracy. There are practical things that might be done to reduce a national population; for example: compulsory sex education, free contraception, abortion (and euthanasia) on demand, financial penalties for too many children, and (rather drastically) a deliberate lessening of the care given to premature babies. As all of these have their difficulties, you can see why education is the usual response.

So will the world cope? Will the targets around ending hunger and malnutrition in the second Sustainable Development Goal be met, even as the population continues to rise, and the climate changes? Well, many argue that even though the population is not necessarily growing beyond food production, as Malthus predicted, it is completely dependent on our continuing use of oil (for fertiliser and energy), and so we will have to develop competitive alternative sources of energy and adequate water supplies before we can expect a growing population to not be a concern. The more mundane matter of eliminating existing wastage ought to be easier to tackle. The UN reckons that currently, around 30% of food is lost during or after harvest through poor agricultural practice, inadequate storage and transport problems, all of which make the produce more vulnerable to pests and wastage. In rich countries, a lot of food is also wasted after purchase.

But it's not just oil and water supplies that are in play here. Soil is being eroded, fish stocks are threatened, biodiversity is being lost, and billions more people are aspiring to a middle-class lifestyle. Perhaps the key factor will be whether we are able to sort all this out in time. The ghost of Malthus, meanwhile, is waiting in the wings to say, "I told you so!"

Notes

1 *The Gathering Storm in Jane Austen's Time* explores the debate between Malthus and Godwin, and includes details of the calculations that persuaded Malthus that we faced population problems. Web link: ow.ly/7BN0308S2Ln.
2 World Bank data on cereal yields are online. Web link: ow.ly/1tDA309drGZ.
3 Sen, A. (1999). *Development as Freedom*. Oxford: Oxford University Press.
4 The *About Geography* blog has a commentary on the world's population. Web link: ow.ly/yh1c308S2yl.

42

LIVING WITHIN LIMITS

The Brundtland Report, *Our Common Future*, described sustainable development like this: "development that meets the needs of the present without compromising the ability of future generations to meet their own needs."[1] The report said that meeting the needs of the world's poor should be a priority, adding that doing this was difficult because of the lack of appropriate social organisation and technology. In this sense, the idea of sustainable development contains an ethical commitment to the well-being of everyone on Earth, as well as to the biosphere as our natural life support system. Thus, sustainable development is not really about the environment; rather, it's about our capacity to change how we live so that we don't upset the delicate balance between ourselves and the biosphere.

In 2012, Kate Raworth, working with Oxfam, captured this idea in a striking fashion, with a way of thinking about sustainable development that is completely different from the way that classical economics sees it. It combines the concepts of planetary boundaries and social ones.[2]

Raworth said that achieving sustainable development means ensuring two things:

(i) All people have the resources needed – such as food, water, healthcare and energy – to fulfil their human rights.
(ii) Our use of natural resources does not put critical biosphere processes under stress.

She then created a framework to capture these ideas. In this, a social foundation forms an inner boundary, and an environmental ceiling forms an

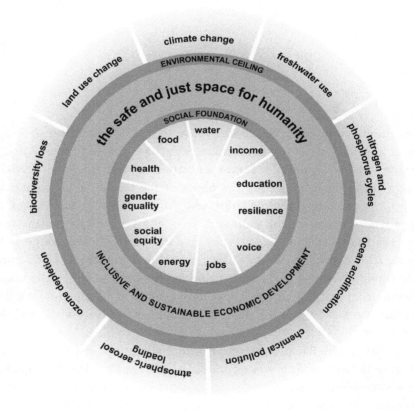

Figure 42.1 A safe and just space for humanity to thrive in showing the environmental ceiling and the social foundation

Source: Oxfam

outer one. The social foundation is based on ideas that gave rise to the Sustainable Development Goals, and the environmental ceiling is based on the planetary boundaries set out by Rockström et al. in 2009.[3] Between these boundaries, there's a doughnut-shaped space that represents the environmentally safe and socially just place where humans can thrive in an economy that doesn't damage the biosphere.

This is a compelling image that has much to recommend it as a way of representing the scope of the problem. It shows 11 aspects of the social foundation (listed on the inside), which are things, Oxfam and Raworth say, that are crucial for a happy and fulfilled life. It also shows nine aspects to the environment (listed on the outside) that represent the essentials of a well-functioning biosphere. Currently, there are many problems in these environmental aspects, and many of these are global in nature, although they can vary in severity from place to place. Similarly, for many people across the world, many of the social aspects are negative features in their lives. There is huge variation in this not only between countries, but also within them.

There are, however, some problems with this way of portraying these ideas, which illustrate some of the difficulties of thinking about sustainable development. Most fundamentally, these are about the idea of a boundary. The social foundation boundary is the minimum entitlement for a decent life, and the environmental ceiling is the boundary beyond which very significant environmental consequences are likely. But these social and environmental boundaries are not the same sort of boundary, and we think about them in different ways.

The atmospheric boundary is physical in nature, and it embodies real limits. For example, there will be critical natural thresholds for ocean acidity before marine ecosystems are damaged beyond repair, and a proportion of atmospheric carbon dioxide beyond which positive feedback loops develop, which mean that temperature rises no matter what we do. Currently, for example, the United Nations says keeping atmospheric CO_2 below 350 ppm is our best guesstimate at staying well below the critical failure limit. The important point is that as a human society, although we can learn something about the nature of such limits, we cannot decide for ourselves what they are.

This is not the case with the social foundation boundary, which is more likely to be relative than absolute, and amenable to social construction and decision-making. For example, if income poverty (defined in 2008 by the

World Bank as <$1.25 per day) looked as if it might be eradicated, the level would have to be revised upwards in order to focus attention on helping the world's poorest people. This has now happened when the World Bank set a rate of $1.90 in 2015 because of the success of the Millennium Development Goals.[4] In a similar way, the level of a national minimum wage can be adjusted through political decision-making, as can taxes, to encourage or discourage particular consumption, or policies on the preferred balance between renewable and non-renewable energy.

In this sense, one boundary is more fundamental than the other, and we do ourselves no favours by saying otherwise. That does not mean that one boundary is more important than the other, and international attempts to do something about the problems reflect this. For example, the 2015 UN Paris Agreement on climate change has a clear focus on how our economies affect the biosphere, as did the 1987 Montreal Protocol. Similarly, the 2016 Sustainable Development Goals (SDGs), with their targets, is evidence of an attempt not just to meet the needs of the world's poor, but to help everyone live a life they have reason to value.

Notes

1 The Brundtland Report, Our Common Future. Web link: ow.ly/lv78308LZOw.
2 Raworth, K. (2012). *A Safe and Just Space for Humanity: Can We Live Within the Doughnut?* Oxford: Oxfam. Web link: ow.ly/NkRF308Moof.
3 Rockström, J. (2009). A safe and just space for humanity. *Nature*, 461, 23 September. Web link: ow.ly/JEov308MobK.
4 *The Guardian* (2015). World Bank: "extreme poverty" to fall below 10% of world population for first time. 5 October. Web link: ow.ly/IfiM30at7AS.

43

ENERGY POLICY

Every country has an energy policy – something that directs how the energy that it uses within its economy is organised, and how it will develop over time. The more top-down and centrally directed a country is, the tighter and more directed the energy policy will be. Those countries with a more laissez-faire approach to their economies will be more likely to leave at least some of the policy to the market. Either way, energy policies are always political in that they are shaped and changed by politics and politicians, sometimes dramatically so, as the case of Germany illustrates.

Although almost 25% of its electricity came from nuclear power, in May 2011 the German coalition government took a sudden decision to close its nuclear power stations by 2022. This followed the Fukushima incident in Japan in March of that year when a number of coastal nuclear power stations were inundated and disabled by a tsunami. Germany has long had a vocal opposition to nuclear power that didn't just come from its strong green movement, and it experienced mass anti-nuclear protests in the wake of Fukushima. Fred Pearce, an environmental commentator, says that German environmental concerns about energy are not just about CO_2 emissions, and that, if anything, Germans worry more about nuclear power than they do about climate change.[1] The German government decision was in response to such attitudes. The seven oldest reactors were immediately taken offline for a safety review, and will never be used again, and an eighth plant, which was already offline, has been shut down as well. Six others will go by 2021 and the remaining three by 2022. Between them, they still (in 2016) provided 15% of German electricity.[2]

In this policy of *energiewende* – energy transition – the phase-out of nuclear was set alongside a target to cut carbon emissions by around 80% by 2050 (from 1990 levels). Under *energiewende*, the plan is that wind and solar, and other renewables such as biomass, hydro and geothermal, will supply 80% of Germany's electricity and 60% of its total energy.

Whether you see this as a panic measure that had more to do with coalition government politics than safety, or as a welcome return to rational thinking about power in response to climate change, the decision left the country with an unwanted problem – how to replace the nuclear electricity. Inevitably, not all will be replaced as it is hoped that efficiency savings will result in a 10% cut to electricity use over the next decade. The intention is also to increase the amount of wind energy that is generated. The problem here is that wind energy is produced in the north of the country, with many wind farms in the North Sea, but a lot of the nuclear plants are in the south. As a result, there are now new protests from those living in heavily forested central Germany about a north-south *energie autobahn* of pylons and very high-voltage DC power lines to carry electricity south to the industries that need it.

Meanwhile, lignite (brown coal) is plugging the immediate gap, and in 2015 contributed around 25% of Germany's electricity, more than solar and wind combined. Lignite is a particularly polluting kind of fuel as its carbon content is usually <50%, and it is very good at generating air pollution. All this has thrown German attempts to cut carbon emissions into reverse, at least for the time being. Fred Pearce[3] quotes Carel Carlowitz Mohn of the European Climate Foundation as saying that lignite is "the blind spot of *energiewende*." One reason is that lignite is cheap and mines located in the relatively poor, ex-communist east of the country provide much-needed jobs. As there are at least 50 years of lignite reserves, mining it may be a rich source of employment, carbon dioxide and air pollution for some time to come.

Any country that looks to use renewables has to have a policy to fill the gaps when the *intermittent* nature of renewables means that the supply falters. That is, when the sun does not shine enough (or at all), and the wind does not blow at the right speed. Neither nuclear nor coal are good for this as power stations based on these fuels take a long time to fire up, and so it makes no sense to keep shutting them down. In fact, with both lignite and nuclear, this is unwise, although for different reasons. That does not rule out some nuclear in an energy mix – it is, after all, carbon-free – unless you think it is inherently dangerous and/or the work of the devil. The best option for the Germans would be natural gas, providing there is carbon capture and storage, as burning natural gas produces about 40% less CO_2 than any form of coal, and it is much cleaner than lignite. Moreover, modern gas turbines can be switched on (and off) in about 10 minutes, making gas ideal as a gap-filling fuel.

But for the Germans, gas is more expensive than lignite because of the failure of the EU's carbon cap and trade system to work properly, and a lot of the gas (maybe a third) comes from Russia, a country that has used gas supply as a political weapon in the past. To complicate things further, the Germans also have a strong dislike of the idea of carbon capture and storage. As we note elsewhere, when continent-wide smart grids are in place, the problem of how to provide enough and consistent electricity from renewables is likely to be solved, but in the meantime dirty German lignite looks set to be the replacement for its nuclear electricity for quite some time, and it is hard to see that this was an outcome that anyone would have planned.

There is an Energy Charts website that shows the fuels used in German electricity production, and how this varies from hour to hour and day to day.[4] This also shows the amount of electricity that Germany imports. As a lot of this comes from France, which produces around 75% of its electricity from nuclear power, which shows you can refuse to bake your nuclear cake and have it too.

The Germans, of course, are not alone in having these problems, and examples such as this could be found in other countries. The UK, for example, has two main energy policy goals:

• to put the UK on a path to cut carbon dioxide emissions by some 60% by about 2050 (from 1990 levels); and
• to maintain reliable energy supplies.

The first of these relates to total energy use, and the second to electricity. Not everyone is confident that both can be achieved at the same time. For example, a British newspaper story[5] in 2015 began: "The UK's energy policy is an act of national suicide – Britain is heading for the greatest self-inflicted political disaster in our history." Another headline in the spring of 2016, in a paper[6] at the opposite end of the political spectrum, said: "UK energy policy is in disarray." Both these were commenting on the lack of security of supply. The first was a critique of the plan to take the carbon out of electricity production; the second was concerned by the prospect of winter power cuts because of the lack of reliable supply, and by the cost of the steps necessary to prevent this. It seems it is harder to please everyone than it is to keep the lights on.

Notes

1 Fred Pearce's Yale Environment 360 blog: "On the Road to Green Energy, Germany Detours on Dirty Coal." Web link: ow.ly/wdRf308S3zs.

2 The European Nuclear Society website has details of the nuclear power stations in Germany. Web link: ow.ly/sWXI308S3Hg.

3 Yale Environment 360 blog: "On the Road to Green Energy, Germany Detours on Dirty Coal." Web link: ow.ly/yFSE30g3Gpp.

4 A range of charts showing German energy use (with English text) can be found online. Web link: energy-charts.de.

5 Christopher Booker writing in *The Daily Telegraph* (21 November 2015). Web link: ow.ly/2hfR30g3GFd.

6 Fiona Harvey writing in *The Guardian* (1 March 2016). Web link: ow.ly/m4LF30g3JVJ.

44

CARBON CAPTURE AND STORAGE

Air contains carbon dioxide (CO_2) because of the respiration of plants and animals. In pre-industrial times, its concentration in air was about 0.03% but it is now over 0.04% because of the release of CO_2 from the burning of fossil fuels in industry and for power. This is an increase of around 35%. This extra CO_2 is making an unwelcome contribution to global warming because of the greenhouse effect. Much of the world's energy strategy is now focused around reducing this production of CO_2 in order to keep the average global temperature rise to below 1.5°C. The 2015 Paris Agreement was seen as a big step towards the possibility of achieving that.

Global energy strategy involves switching from fossil fuels to renewables, for example by using solar, wind, hydro, biomass and other renewable sources of energy to generate electricity, and by using electricity or hydrogen to power vehicles. Progress is being made in this, although the transition is slow because there is so much of our existing infrastructure devoted to fossil fuels. Just think how many service stations there are up and down the country selling petrol and diesel, compared to electric vehicle charging points or hydrogen charge stations. Meanwhile, more and more CO_2 is being produced every day, and this will be the case for some time. Because of this, one of the current big ideas about managing this transition to renewables is carbon capture and storage (CCS). If you did much chemistry in school, you might remember an experiment in which air was bubbled through a liquid, which then turned cloudy. The liquid was lime water, and this showed that air contains CO_2. The carbon dioxide reacted with the lime water to form solid calcium carbonate, which is widespread on Earth in chalk and limestone rocks. Thus, the CO_2 was captured and removed from the air.

Carbon capture and storage technology (which was first developed in the 1970s) would be a useful transition strategy, the argument goes, as we could carry on burning fossil fuels without releasing CO_2 into the air. A counterargument to a CCS approach is that it just slows down the transition to carbon-free energy. The fossil fuel industry obviously likes the carbon capture idea; those keen on a swift transition do not. Sadly, lime water itself is no good for carbon capture because in order to make it, CO_2 is produced as well. In fact, as much would be produced as is then removed, which is an expensive way of achieving nothing at all. That said, a chemical process to remove CO_2 and form metal carbonates would be ideal as these are very stable over long periods. There is much research at the moment into ways of doing this that are both efficient and cost-effective.

The most basic form of carbon capture is photosynthesis itself: the process whereby green plants capture the CO_2 in the air and convert it into sugars and other useful materials for the plant. Although the more CO_2 there is in the air, the more photosynthesis there is, it seems that it will not help us remove the extra gas we are now producing through industrial processes.

The idea of removing problem gases from exhaust chimneys is not new. Sulphur dioxide (SO_2) is produced from the burning of coal containing sulphur (sometimes up to 10% by weight). However, this is now routinely removed from power station chimneys by a variety of methods. Although this was first considered over 150 years ago, and initially carried out in power stations in the 1930s, it was only after the problems caused by acid rain in the 1980s that the technology was widely developed and adopted across the world. But none of the techniques that are used to capture SO_2 are much good with carbon dioxide because there is so much CO_2 produced in the combustion of fossil fuels as the proportion of carbon in the fuels is very high. In the best coal (anthracite), it's as high as 90%.

So, how would CCS technology work, and what are the drawbacks? It is possible to remove the carbon from any fuel both before it is burned and after. Both processes are well understood. Doing so beforehand has advantages, and is already widely applied in the fertiliser and chemical industries. Here, a fossil fuel such as oil is converted into a mixture of carbon monoxide (CO) and hydrogen, known as syngas. The CO in the syngas is converted into CO_2 and separated from the hydrogen, which can then be used as the fuel, producing water as the only by-product. When carried out after combustion, the CO_2 gas can be absorbed or separated during its passage up a chimney. However, doing this, and then compressing the gas, reduces the energy efficiency of the production process by at least 25%. Further, adding this technology to existing plants is very expensive, especially if they are a long distance from where the CO_2 is to be stored.

Following its capture, the CO_2 can be piped under pressure into deep holes in the ground. Here, the idea is that the pressures are high enough to keep it from leaking into the atmosphere. Old oil wells are said to be particularly suitable for this. Indeed, high-pressure carbon dioxide is already pumped into wells that are nearing the end of their normal production in order to get a bit more oil out.

In 2016, the Oxburgh Report to the UK government concluded that CCS was an essential technology if the UK is to meet its international agreements in relation to climate change (for example, the Paris Agreement) at minimum cost.[1] It argued that there was ample, safe and secure CO_2 storage capacity in rocks deep beneath UK territorial waters, and that this is the lowest-cost form of storage. It also said that CCS has the potential to safely store 15% of current UK CO_2 emissions by 2030, and up to 40% by 2050. It's clear from these figures, however, that CCS only makes sense if there is also a reduction in the amount of CO_2 produced.

As post-combustion processes are expensive to build, especially if they are retrofitted to existing plants, the ideal is to begin by fitting them onto the biggest emitters of CO_2. But even when all these have been dealt with, that will still leave all the small CO_2 producers such as gas boilers in houses and offices, and vehicles of all kinds. This is another reason why CCS will have to be seen as part of a longer-term strategy of removing carbon from energy provision. For example, if our plan is to stop producing CO_2 altogether, then all the gas boilers and internal combustion engines will need to be replaced with renewable energy sources such as heat pumps, or electricity generated from renewables. And all this is before we begin to contemplate, as we might one day have to, the removal of CO_2 that's already in the air to return its concentration to about what it was in pre-industrial times: around 0.03%.

Note

1 The Oxburgh Report can be found on the Scottish Carbon Storage and Capture website. Web link: sccs.org.uk. Type "Oxburgh" into the search box.

45

ECOLOGICAL FOOTPRINTS AND HANDPRINTS

The idea of the ecological footprint dates from the 1990s when two Canadian academics put forward the metaphor as a means of bringing home to people the impact that they were having on the ability of the Earth's biosphere to support life. It's a clever idea because we all walk on the Earth every day, and the size of our tread – our impact – varies enormously across the globe: from country to country, and from person to person within a country.

The Global Footprint Network says that an ecological footprint is a comparison of the demands we make on nature and its capacity to meet such demands through biological productivity (within forests, oceans, farms, etc.). In this, the size of the footprint (in hectares) represents the productive area required to provide for personal and social living.[1] It's a measure of how fast we are using resources up, compared to how quickly the Earth can regenerate them. It seems clear that since the 1970s, we have been exceeding what the Earth can regenerate, and it now takes the biosphere about 18 months to renew what humans use every year. This led the New Economics Foundation, a UK think tank, to the idea of Earth Overshoot

Day: the date in the year when we begin to live beyond the Earth's means.[2] This gets earlier each year; in 2006, the day fell in October; by 2016, it was on 8 August.

Whoever we are, we must make demands on the Earth in order to live. At its most basic, we breathe, we eat, we need water and we excrete waste. We also need shelter and clothing, the means to socialise and protect ourselves, and to be able to find meaning in life. All humans do all of these, one way or another, but there is a huge difference in the levels of such activity across the globe: from isolated societies still untouched by modernity, to subsistence farming that may be fruitful or hand to mouth, to poor communities in sprawling townships, favelas and shanti, to the middle classes, and onto the elite who criss-cross the planet on private jets from one glamorous location to the next.

Put like this, it's easy to see how impact varies, but the idea of the footprint went farther by putting numbers onto it, and there are now numerous online calculators that allow you to estimate what your personal or family impact is. Typically, these will look at your living arrangements, travel, house heating, and the sort of food you eat. Some will ask you for actual data (per annum miles travelled by air and actual heating costs, for example); others will offer you a range to choose from. Typically, the footprint is given in hectares, which is a rough measure of how much land on the Earth is being used to provide your demands upon it. One calculator, for example, gives responses in this form: "Your ecological footprint is estimated at 6.4 global hectares, and if everyone lived like you, we'd need 3.6 planets to support global consumption." Here, the 3.6 planets figure is a measure of how excessive the software designers think your lifestyle is. Some of these measures are pretty crude, and it's easy to get a low score by lying about how many long-haul flights you take, or denying that you're from the US.

Often, these calculators will give you a measure of your carbon footprint (in tonnes) as well.[3] This is a measure of how much carbon dioxide your activities are dumping into the atmosphere each year. Unsurprisingly, perhaps, the calculators provided by environmental organisations come with a bit of behavioural moralising on the side. Here's what WWF says on its website: "Our lifestyle choices make up our environmental carbon footprint. Measuring yours takes less than 5 minutes and could change the way you live." When you take part, each step is accompanied by useful information and hints about changing how you live – for example:

Buying local food will generally (though not necessarily) result in a lower footprint. It depends how it's produced: if grown in natural sunlight, in season, with little artificial fertiliser or pesticide, and not over-packaged, it will have a good environmental footprint. And buying from abroad needn't always be a bad thing – food grown in a non-intensive, sustainable way, and traded fairly, can also provide vital income for developing countries – but the food miles will contribute to your footprint. Buy thoughtfully.

It's probably fair to say that unless you're a hermit, no one in a developed economy will get very low ecological or carbon footprint scores, and so there's always room for improvement; that's to say, opportunity to live and tread more lightly on the Earth. If you're very poor, however, getting low scores is much more straightforward.

There's something of a consensus among development-focused charities that the world's poor should have higher ecological footprints than they currently do, as these are proxies for a decent standard of living, and that the world's rich should have lower ones. The latter is something rich-world politicians find hard to put into policy, let alone practice. The Sustainable Development Goals are one means of helping the first of these to happen, without leading to an increase in our global carbon footprint. This is going to be a neat trick if the UN, governments and international agencies manage to pull it off.

One problem with the ecological footprint idea is its negativity in that there's a relentless focus on reducing it – on living more lightly on the Earth – because of what we are doing wrong. This has led some, such as Daniel Goleman, to argue that doing this engages the wrong part of people's brains, and that we need to focus on more positive things. He believes we would generate more positive and enthusiastic responses by talking instead about what's termed an environmental handprint.[4] This is a measure of what we are doing right; for example, cycling, eating locally produced food and using renewable energy.

Goleman says that such positive psychology (we'd always be trying to increase our handprint) could encourage communities to drive sustainability as an idea. Some see footprints and handprints as complementary, and that the aim ought to be to have a larger handprint than footprint. Of course, the problem with this approach is that it's just the same idea as carbon offsetting – where you do good works (tree-planting or compost-ing, say) to make up for the fact that you're jetting around the world on holiday or eating a lot of meat. While it sounds very much like atoning for your sins, and is certainly net-positive, some think that it would be better for you, and for everyone else, if you just sinned less often in the first place.

Notes

1 The Global Footprint Network explains the ecological footprint. Web link: ow.ly/MjqZ308PUFg.
2 Earth Overshoot Day. Web link: overshootday.org.
3 Ecological Footprint calculators can be found online. Web links: ecological footprint.com and footprint.wwf.org.uk.
4 Handprinting is explained by the Harvard TH Chan School of Public Health. Web link: ow.ly/CRaO308PV3X.

46

THE CIRCULAR ECONOMY

The idea at the heart of the circular economy is not a new one.[1] Terms such as cradle to cradle have been used for a while to describe an industrial economy that sets out not just to reduce and minimise waste, but to use whatever waste is created to produce new products. Cradle to cradle stands in sharp contrast to the better-known phrase cradle to grave, which more aptly describes our current economy, as well as the fate of us all.

Our current economy is a linear one that is often characterised in this way:

Take   →   Make   →   Dispose

Figure 46.1 A characterisation of our current (linear) economy

Source: The Ellen MacArthur Foundation

Here, waste is the dominant feature. Raw materials are taken out of the earth and converted into useful materials, which are then fabricated into a product that can be sold to consumers. The consumer eventually tires of the product, or wears it out, and then disposes of it – sometimes after a bit of recycling – into landfill. At each of these stages, waste is generated, and usually also discarded, often polluting land and water. Recycling only delays the inevitable. At each of the stages, energy is also used in the conversion process, sometimes significant amounts of it. Since the Industrial Revolution, this energy has mostly come from fossil fuels mined out of the earth, and still largely does so despite their use being identified as the major cause of climate change.

This linear industrial model contrasts poorly with how biological systems operate in nature. These are always circular as materials and energy flow round loops, and there is no material waste as it is always consumed by another organism. All these (biochemical) processes are low-temperature ones, and all the energy used is renewable as it comes from the sun. Here, the key idea is that waste = food.

A linear economy made some sense when resource materials were cheap because there was little demand for them, or they were abundant, or pollution could be contained. This has been the case, more or less, up to quite recently. But now it's different. Some raw material prices are rising in real terms because there is increasing demand *and* because they are becoming scarcer. The rising demand is both technological and demographic. For example, elements such as neodymium, terbium, yttrium and dysprosium are now found in a wide range of products such as TV screens, wind turbine magnets, mobile phones, cameras, computer hard disks, lithium batteries – and increasingly in almost everything: think smart fridges and even smart walking sticks. The ores of these rare earth elements are often in hard-to-reach places, some of which are politically unstable, all of which adds to price uncertainty. At the same time, the demand for such products is increasing as people in all the world's economies – advanced and developing – see their usefulness and recognise their value. A recent World Bank report said that in 2012, there were already 650 million mobile phone subscribers in Africa alone. Such changes have made a number of companies think again about the linear economy and how it can be changed in the face of these developments.

The key to this thinking is new ways of designing products so that they can be "made to be made again." In other words, making products that are easy to take apart so that valuable materials can be recovered and reused. This tackles head-on a key problem that has locked the economy into a linear mode: the fact that products have been made in such a complex way that the only feasible thing to do at the end of their useful life has been to throw them away. But this is changing. Industrial carpet tiles, which can be completely taken apart at the end of their useful life, are now established in the market. At the end of their use, the company takes them back and creates new tiles; this process even includes unmaking and remaking the polymers that are used in them.

Another key facet of the linear economy has been the tendency of companies to give away the ownership of their products, and all the materials within them. This may seem an odd way to put it when what they actually do is sell them, but it amounts to the same thing. In contrast to this, there are now novel ways of turning products into services. For example, Phillips, a Dutch multinational company, will now sell businesses light, rather than light fittings and bulbs. You can have a contract with the company that guarantees you so many lumens per square metre of light for a particular length of time. Phillips puts in the fittings and pays the electricity costs. All you do is pay for the light you use. The materials, both rare and everyday, remain the property of the company. This approach may well soon extend into the white goods area of washing machines and dishwashers. The argument goes, people don't need to own such machines; what they really need are clean clothes and dishes. The idea here will be to guarantee (and sell) so many wash cycles. The consumer gets what they want without owning (or consuming) the products. These can be refurbished by the company, which keeps ownership of the materials. This is already commonplace in the photocopying industry. Whether it will catch on in the domestic market remains to be seen, as we are all so addicted to the idea of owning our own stuff.

The idea of the circular economy has been promoted by the Ellen MacArthur Foundation (EMF), which has demonstrated convincingly that such an approach would be much more efficient in terms of resource use, profitable for the companies involved, and beneficial for the consumer. A study by McKinsey, an international consultancy, argues that by adopting circular economy principles, the European economy would benefit by €1.8 trillion by 2030 – 50% more than by following a linear development path.[2] This would result in an increase of household incomes, and a halving of carbon dioxide emissions.

The foundation's approach has been to identify what they term biological cycles and nutrients, and technical cycles and nutrients, where keeping these separate is a key objective. This means keeping anything made out of mineral raw materials away from living systems, as this EMF graphic shows:

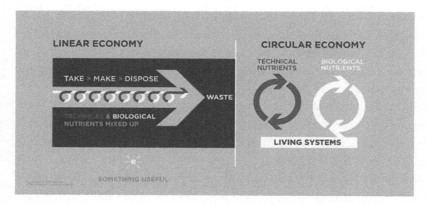

Figure 46.2 A comparison of the linear and circular economies

Source: The Ellen MacArthur Foundation

A circular economy aims to maintain products, components and materials at their highest utility and value. It will have a cycling of materials, and an efficient cascading of renewable energy through the system, and it will restore or regenerate natural and social capital. Its key feature, however, is that economic growth and development can be decoupled from the consumption of finite resources.[3] If successful, this would be a major achievement.

Notes

1 A set of circular economy principles can be found on the Ellen Macarthur Foundation website. Web link: ow.ly/y1bu308S4bU.

2 McKinsey's (2016) report, *The Circular Economy: Moving from Theory to Practice*, is a collection of articles about the transition taking place as companies use circular economy concepts to capture more value from resources and to provide customers with better experiences. Web link: ow.ly/pWkZ308S3Tj.

3 The Ellen MacArthur Foundation's *Growth Within: A Circular Economy Vision for a Competitive Europe* is in the "Books and Reports" section of their website. Web link: ellenmacarthurfoundation.org.

47

THE EARTH CHARTER

Have you heard of the Earth Charter?[1] Perhaps not, as it rarely features in the news or in what governments have to say. In a way, this is surprising, as it's an international declaration of fundamental values and principles that its creators and supporters say are now needed to build a just, sustainable and peaceful society across our planet.

This is the preamble to the Charter:

> We stand at a critical moment in Earth's history, a time when humanity must choose its future. As the world becomes increasingly interdependent and fragile, the future at once holds great peril and great promise. To move forward we must recognise that in the midst of a magnificent diversity of cultures and life forms we are one human family and one Earth community with a common destiny. We must join together to bring forth a sustainable global society founded on respect for nature, universal human rights, economic justice, and a culture of peace. Towards this end, it is imperative that we, the peoples of Earth, declare our responsibility to one another, to the greater community of life, and to future generations.

Although the language is rather flowery, and it lacks detail about the troubles we all face, the key idea is straightforward: people should come together to create a sustainable world based on respecting nature, peaceful coexistence, and human rights and justice for all. In reality, that means governments coming together, which is perhaps why it has not been very successful in achieving its ends.

In a way, it might be seen as the Magna Carta for our times as that great charter was also concerned with rights, responsibilities and justice, although the English barons that forced King John to sign it were more concerned with their own interests (rather than the people's), and were certainly not as troubled as we are by environmental issues on a global scale. In part, this was because humans had not yet had the opportunity to properly mess with the biosphere, but it was also because it was hard enough to know what was happening across the country, let alone the world. Our great skill at communication these days means we now know a great deal about the difficulties the globe faces, and although that can sometimes help us fix them, that doesn't always work out.

The idea of the Earth Charter originated through the United Nations when the World Commission on Environment and Development (the Brundtland Commission) called for a means of guiding the transition to sustainable development. However, nothing came of this at the time. It took off in 1994 when Maurice Strong and Mikhail Gorbachev revived the idea as an initiative outside the framework of the UN and used a global consultation process to write a draft charter. The final version of the Charter was agreed in 2000, and then began the process of promoting its dissemination, endorsement and implementation by civil society, business and government. An attempt to secure endorsement in 2002 at the UN's World Summit on Sustainable Development in Johannesburg failed, and formal recognition by the UN has still not happened.[2] Given the recent creation of the Sustainable Development Goals, which are similar to but not the same as the goals of the Earth Charter, perhaps it never will be.

The Charter has four pillars. These are:

1   *Respect and care for the community of life*, which includes the interdependence of, and the need to respect, "all life," the need to guarantee human rights and fundamental freedoms, and the responsibility to use resources responsibly.

There's a problem with this sort of language. For example, how many of us would agree that respect for life forms such as the Norwalk virus and other versions of norovirus should not stop us from trying to eliminate them? There have to be limits, and the Charter isn't good at drawing them. Similarly, not everyone sees "human rights and fundamental freedoms" in the same way, particularly perhaps where religious perspectives and cultural traditions are brought to bear. The Charter makes a lot of assumptions about all of this.

2   *Ecological integrity*, which includes an emphasis on bolstering biodiversity and protecting the natural processes that sustain life, on the careful use of resources, and on always erring on the side of caution when it comes to the possibility of environmental harm from economic development.

There's an emphasis on recycling, but no mention of the circular economy, with its focus on careful product design to maintain the quality of valuable resource stocks, and this betrays the Charter's age. Though many will likely agree with its emphasis on the negative impacts of GMOs and non-native species, there are also many who think that without the technology of genetic modification, the world's hungry will not be fed. And there are now ecologists saying that we should stop obsessing about non-native species as these are likely to be needed if we're to adapt to climate change.

3   *Social and economic justice*, which includes eradicating poverty as an ethical imperative, and a wide-ranging emphasis on equality and equity, with a particular emphasis on: (i) the life chances of girls and women because of the continuing widespread discrimination and violence against them; and (ii) the rights of indigenous peoples.

This is a wonderful wish list, and many will sign up to it; what's missing, however, is a sense of realism and any insight as to how it's going to happen. It glosses over the awkward fact that such social progress is often the result of trade-offs between interests and preferences, and it perpetuates the myth that there will only be winners from the sustainable development process.

4   *Democracy, non-violence and peace*, which includes strengthening democracy, making education relevant to a sustainable world, promoting a culture of tolerance, non-violence and peace, and treating all living beings with respect, in the sense of preventing cruelty and suffering to animals.

Its aim to eliminate all nuclear, biological and toxic weapons is surely something we might all agree on, but perhaps not the emphasis on demilitarisation to the level of a "non-provocative defence posture," and converting military resources to peaceful purposes. Although many countries have already taken strides in this direction, perhaps the precautionary principle has to apply here as well, and we'll give the last word to Hillaire Belloc's cautionary tale, where "Pale Ebenezer thought it wrong to fight, but Roaring Bill, who killed him, thought it right."[3]

Eight hundred years on, the influence and impact of Magna Carta are clear, although they are not yet universal. We wonder whether it will take as long for the values embodied in the Earth Charter to become accepted, or whether they will ever be, because of the contradictions they embody and the assumptions the Charter makes about the perfectibility of humanity.

Notes

1 The Earth Charter. Web link: earthcharter.org.
2 This is a timeline of the Earth Charter's development. Web link: ow.ly/vRS3308PUqU.
3 Belloc, H. (1923/1938). *Sonnets and Verse*. London: Duckworth.

48

BREXIT AND
ENVIRONMENTAL LAW

We are writing this early in 2017 before Article 50 of the Lisbon Treaty has been triggered, and in advance of any negotiations about Brexit – that is, the separation of the European Union (EU) and the UK. All we know, therefore, is that this means leaving the EU, whatever that will turn out to mean in reality. As such, this chapter is not about the politics of the process, or its desirability, but about how a separation might affect environmental and sustainability policies and issues.

In the build-up to the 23 June referendum in 2016, there was very little focus on environmental or sustainability issues, at least at the national level, and yet the implications for the UK's environment could be far-reaching. The Wildlife Trusts say that 80% of UK environmental law comes from the EU, which has a large number of environmental policies that have been built up since the 1970s, and all apply to the UK. These cover issues such as air and water pollution, land use, agriculture and fishing, renewable energy targets, waste reduction, protection of threatened species and habitats, planning, and climate change. These policies are said to comprise the most comprehensive set of environmental protection law ever devised. Now that there has been a vote to leave, it remains unclear whether environmental standards will stay the same, rise or fall.

In the short term, it looks as if all EU environmental law will have to be incorporated into British law, and then decisions made about which parts to keep, which to abandon, and which to amend, but it's unclear how that will play out. Immediately after the vote, there was much angst (if that's the right word in these circumstances) among those who speak for wildlife and the biosphere more generally. There was a fear that all the positive features of existing EU law would somehow be abandoned as soon as possible, if not actually overnight. Quickly, however, it was appreciated that this would not happen. This was rapidly followed by the realisation that this rupture presents an opportunity to forge a better settlement between farming, conservation, rural communities and the environment more generally. In other words: a chance to rethink land-related payments and subsidies of all kinds, and to forge new policy. Such policy might, on the one hand, make agriculture fit for wildlife and for rural working and living, and, on the other, mean that farmers are enabled to grow nutritious food and other crops for the UK's needs in a sustainable way; sustainable, that is, in terms of both the environment, and their own business development. As the Country Landowners Association noted: "Brexit has given us the opportunity to develop a new food, farming and environmental policy which can deliver even greater benefits for the natural world." At the same time, a YouGov poll reported that UK adults think Britain should either have the same (38%) or stronger (42%) environmental protection after Brexit.

A significant part of the EU influence on UK environmental policy and practice comes through the Common Agricultural Policy (CAP) and the direct cash payments to landowners that influence what goes on in the countryside.[1] These payments amounted to some £3.2 billion in 2014–2015.[2] Many landowners get significant funds that are classed as either rural development, market schemes or direct aid. Not all recipients are farmers, as the following examples illustrate:

Table 48.1 Examples of Common Agricultural Policy payments

Norfolk Wildlife Trust	£1,007,375
Cornish Scrumpy Co.	£1,180,689
Ballindalloch Distillery	£1,273,123
Frank A. Smart & Son Ltd	£2,986,506
RSPB Scotland	£3,584,032
Berry Gardens Growers	£4,528,815
Welsh Assembly Government	£7,671,463
National Trust	£7,779,991

Source: Defra

Many such payments cause eyebrows to be raised, and among the more controversial was the £405,960 paid to Juddmonte Farms, which Greenpeace says is the racehorse-breeding operation of Prince Khalid Abdullah of Saudi Arabia.[3]

A complex web of regulations guides these payments. Some seem straightforward. For example, Berry Garden Growers received funding as "aid in fruit and vegetables sector," but the complexity of the scheme is shown in the detail of the payments to RSPB Scotland. These, which we don't really understand, include:

Table 48.2 Common Agricultural Policy payments to RSPB Scotland

Single area payment scheme	£1,961,450
Agri-environment-climate	£773,279
Forest area development and improvement of the viability of forests	£484,145
Forest environmental and climate services and forest conservation	£175,721
Payments to areas facing natural and other specific constraints	£86,351
Investments in physical assets	£54,634
Animal welfare	£1,525

Source: Defra

The Treasury has guaranteed direct payments for land ownership until 2020, although it has not said it will continue to fund wildlife protection on farms. Meanwhile, Defra (the Department for Environment, Farming and Rural Affairs) is working on a 25-year plan to protect the environment. In late 2016, the President of the Wildlife Trusts, Tony Juniper, argued that there are five policies, rules and laws that we must try to hang on to as they guide much of how Britain approaches conservation and environmental challenges.[4] These are:

- nature protection rules under the Birds and Habitats Directives that protect our most cherished wildlife and special natural places on land and at sea;
- policies that govern everything from the state of rivers to the quality of the air;
- CAP rules geared to meeting ecological goals in farmed landscapes;
- Common Fisheries Policy rules requiring the sustainable management of fish stocks; and
- EU agreements to combat climate change.

Juniper said that incorporating these into UK law will not be enough, however, and that Defra's 25-year plan should build a legal framework around them that can survive changes of government. The 2008 Climate Change Act has done this by setting a long-term goal, with strategy being evolved by government departments and advisory committees as time passes and circumstances change. Juniper argues that we should do all that in relation to the environment more generally, setting long-term goals for habitats and species, and aligning nature protection with climate change goals.

To keep pressure on government during Brexit negotiations, 13 of the UK's main environmental organisations, with a combined (though overlapping) membership of 7.9 million people, have formed Greener UK.[5] This is in order to use the Brexit negotiations to make the UK a world leader on the environment, and MPs are being urged to support this through the Green Alliance pledge.[6] Early in 2017, the influential House of Commons Environmental Audit Committee released a report saying that a new Environmental Protection Act should be passed before the UK leaves the EU, providing equivalent or better protection for the environment than before.

Notes

1 CAP Guidance notes from the Countryside Landowners Association. Web link: ow.ly/gznH308SZ1H.

2 A list of who gets the UK's CAP payments can be found online. Web link: ow.ly/IOkm308SYKV.

3 Greenpeace's report of its investigation of the link between the Common Agricultural Policy (CAP) payments and tax havens is online. Web link: ow.ly/ZX3y308SYg4.

4 Tony Juniper's blog can be found online. Web link: ow.ly/8z3y308SY8P.

5 Greener UK is a group of 13 major environmental organisations, with a combined public membership of 7.9 million, united in the belief that leaving the EU is a pivotal moment to restore and enhance the UK's environment. Web link: ow.ly/vTko308SXZm.

6 The Green Alliance is an independent think tank that works with businesses, NGOs and political parties to stimulate new thinking and dialogue on environmental policy. Web link: green-alliance.org.uk.

49

PROTECTED AREAS

Luc Hoffman, heir to the Hoffman-Roche pharmaceutical empire, was a committed conservationist and a key player in the formation of WWF, but he was always troubled by the idea of nature reserves.[1] Ideally, for Hoffman, the long-term future of our planet's biodiversity lay in achieving a form of global human development that could exist in harmony with nature so that special protection in reserves became unnecessary. So when we learn that between 1990 and 2014, the total number of designated protected areas in the world rose from 13.4 million to 32 million square kilometres, we have to ask ourselves whether this is, incontrovertibly, a good sign.[2,3]

By 2016, over 15% of the world's land area and 3.4% of the oceans were protected, with every country having a protected area system. This includes protection for plants, animals, geology, cultural heritage, and whole ecological communities and processes. For some, this is a signal that we are getting better at protecting the fragile ecosystem that supports life on Earth. For others, these areas are glorified zoos and botanical gardens that, even as they are defined, allow the rest of the planet to be despoiled unhindered. What's more, areas that are protected for future generations can fall victim to pressing human development priorities that undermine them – quite literally, in the case of fossil fuels and other mineral resources. This is either by incremental development nibbling at their edges, or through wholesale deregistration. Worse still, local populations, leading traditional, evidently sustainable lifestyles, often pay the price of state-sponsored protection.

Humans have always set aside special places often for ceremonial or other cultural or religious purposes. This can still be seen today in the case of sacred groves in Ghana, which have been recognised by conservationists as a justification to preserve wild areas. Similarly, in the British countryside, numerous remnant woodlands survived because of their use as coverts that were valued by fox hunting landowners. In fact, many protected forests across Europe remained intact because of their role as hunting grounds of the nobility. As the Industrial Revolution made its mark across Europe, the Romantic movement highlighted the value of the remaining untamed countryside. The poet William Wordsworth even proposed that England's Lake District be seen as: "a sort of national property, in which every man has a right and interest who has an eye to perceive and a heart to enjoy."

Meanwhile, as European colonial powers exploited other parts of the world, they set aside areas for a range of reasons, from watershed management to hunting grounds for elites – normally taking no heed of existing populations. The world's first national park, Yellowstone, designated in the US in 1872, was actually the result of corporate interests, the necessary preparations being largely financed by the Northern Pacific Railroad Company, which saw tourism as a means of growing its business. Other national parks followed in Canada, Australia and South Africa. In each case, Europeans identified tracts of what they termed wilderness, irrespective of the presence of indigenous peoples.

In Europe, most protected areas are lived-in, cultural landscapes. According to the designations of IUCN, these are not national parks at all.[4] The Forest of Dean in England, for example, was a royal hunting forest in the eleventh century CE that was subsequently encroached by mining communities over hundreds of years before being designated a National Forest Park in 1938. It now comprises a mosaic of special areas protected for landscape and biodiversity alongside established urban settlements.

Elsewhere, the process is not always so accommodating. Across Africa, for example, colonial powers evicted people from newly designated national parks, and the process continued beyond independence. Indigenous peoples have suffered evictions by their government officials, often with support from international conservation organisations, for example at

Korup National Park in Cameroon, while in Tanzania a private company assisted in the forced removal of Maasai people from a lion hunting area.[5] In other cases, governments avoid direct expulsions but impose restrictions on traditional practices such as hunting, tree cutting or cattle grazing so that people are forced from their homelands. Examples range from the San of Botswana to the villagers of Borjomi, Georgia; in each case, the logic of sustainable development, as defined by foreign aid-funded projects or corporate interests, wins out over traditional lifestyles.

This need not be an either/or situation, and imaginative efforts have been made to square this circle over recent decades. Joint forest management between local people and state authorities has taken place in Nepal for many years. In Uganda's Mount Elgon national park, shared ownership proved too big a step for the authorities, but a model of collaborative management allows local resource use to be balanced with conservation priorities. Since 1989, Communal Areas Management Programme for Indigenous Resources (CAMPFIRE) has allowed communities in Zimbabwe to benefit from game hunting revenues, another balancing act – in this case, between sustainable resource use (big game hunting) and the sensibilities of the American public whose aid dollars support the programme.[6]

These are IUCN's seven protected areas categories:[7]

Table 49.1 The seven categories of protected areas

1a	**Strict nature reserve:** To protect biodiversity and also possibly geological/geomorphological features; human impacts strictly controlled.
1b	**Wilderness area:** Large unmodified or slightly modified areas, retaining their natural character without significant human habitation.
2	**National park:** Large-scale ecological processes protected allowing compatible, spiritual, scientific, educational, recreational opportunities.
3	**Natural monument or feature:** Protecting a specific natural monument on land or sea.
4	**Habitat/species management area:** Particular species or habitats and management required to protect particular species or to maintain habitats.
5	**Protected landscape/seascape:** Areas of distinct character created by the interaction of people and nature over time.
6	**Protected area with sustainable use of natural resources:** These conserve ecosystems and habitats together with associated cultural values and traditional natural resource management systems.

Source: IUCN

Whatever our view on protected areas, it seems that for many plants and animals to survive, human beings must take steps to protect their habitats, which confers on our species a special responsibility. Yet elsewhere in this book, we have discussed how humans are very much a part of nature and the environment, and this stewardship role appears to be at odds with the notion of our being a biological creature like any other. Then again, doesn't every species adapt its environment simply by being a part of it? In which case, are we not behaving naturally by organising our world in any way that suits our species? Either way, it appears that protected area systems are increasingly being organised around leisure and tourism rather than ecology, as if that's the only way to justify their existence within a largely neoliberal economy.[8]

How we define our place in nature remains a challenge, but while we're working that out, it seems prudent to afford nature as much protection as possible, a conclusion that Luc Hoffman reluctantly came to accept.

Notes

1 *The Economist* (2016). Obituary: Luc Hoffmann. 6 August. Web link: ow.ly/nG5A308S7IJ.
2 The Protected Planet website allows you to search for protected areas around the world. Web link: protectedplanet.net.
3 UNEP's World Conservation Monitoring Centre (WCMC). Web link: ow.ly/ft7q308S8PV.
4 The World Database on Protected Areas (WDPA) is a joint project between the United Nations Environment Programme (UNEP) and IUCN. You can see key statistics on protected areas and endangered species and also share information through various social media platforms. Web link: protectedplanet.net.
5 *The Ecologist* (2015). Tanzania breaks promise – thousands of Maasai evicted to make way for lion hunt. 27 February. Web link: ow.ly/U6Ra308S7xk.
6 Vidal, J. (2016). The tribes paying the brutal price of conservation. *The Guardian*, 28 August. Web link: ow.ly/NUH8308S8eR.
7 IUCN protected area categories classify protected areas according to their management objectives. The categories are recognised by international bodies such as the United Nations and by many national governments as the global standard for defining and recording protected areas. Web link: ow.ly/2zIZ308S7YO.
8 Moran, J.A. (2012). Three challenges for environmental philosophy. *Philosophy Now*. Web link: ow.ly/ykqR308S8pR.

50

RECYCLING

Children are taught at school that recycling is a good thing to do, and that the more stuff that is recycled, the better. But is this really the case?

The collection and recycling of glass bottles and metal cans has been happening for a long time now. Glass waste is usually used to make more bottles and reduce the new raw materials that are needed. Metal waste (mostly steel and aluminium) goes back into the metal production process, again sparing the use of natural resources. Both processes also reduce the amount of energy needed to make new material. It was not so long ago that some glass bottles were collected, steam-cleaned, and reused. Milk brought to the house was in such bottles, and soft drinks used to be in bottles where a returnable deposit had to be paid. Things are mostly different now.

When talking about waste in school, children usually learn a new set of three Rs:

Reduce	Re-use	Recycle

Figure 50.1 The three Rs of the waste hierarchy

Source: WRAP

This is known as the waste hierarchy, and shows you what your priorities should be, particularly about packaging:

(i)   reduce the amount you buy;
(ii) reuse it for something when you've finished with it; and
(iii) recycle it so that other things can be made.

For example, the plastic bags once provided free of charge by supermarkets for purchases could be used to carry the shopping home, then be used again for that, or for storing something at home, and only after that be taken back to the supermarket for recycling.

Some schools go further than this and introduce a fourth R:

| Refuse | Reduce | Re-use | Recycle |

Figure 50.2   Adding a fourth R to the waste hierarchy

Source: WRAP

Here, the idea is to reduce the amount of waste created in the first place, for example by refusing a plastic bag because you've taken your own bags with you, or are using bags for life provided by retailers. Supermarkets were keen on this approach as it saved them money. In 2014, the UK waste reduction charity WRAP says that 8.5 billion thin (single-use) plastic bags were issued by UK supermarkets.[1] This is many fewer than the 12.2 billion that were issued in 2006 when data collection started. In addition to these, a further 0.5 billion reusable bags for life were also issued in 2014. By October 2015, a 5p charge had been introduced on plastic bags provided by most large UK shops. Over time, this nudge is expected to help us use more of our own bags and many fewer new plastic ones.[2]

All local councils take materials for recycling. They are keen to keep them out of the waste stream that goes into landfill sites because the costs of landfill disposal are now so high. Councils are also keen to recover materials with a cash value. Typically, councils have household collections for metal cans, glass bottles and jars, paper and cardboard, clothes, and some plastics; the trouble is that there are over 300 different schemes across the country. Some councils also collect garden waste, and this is turned into compost. Councils may also have recycling centres where materials can be taken. Here, other things will be collected for recycling: light bulbs, oil, metal objects, TVs, white goods, batteries, etc.

The UK recycling rate of household waste increased from 11% in 2000 to 45% in 2014, falling back slightly to 44% in 2015. This suggests that getting to the EU target of 50% by 2020 may not be easy, especially as the amount of waste sorted by householders for recycling, but then rejected and sent to landfill or for incineration, is increasing. Some 338,000 tonnes were rejected in 2014. This is an awful lot of full bin lorries. Councils say that this problem is caused by householders not bothering to separate the waste properly, or not cleaning it well enough before it is recycled. They say the main problems are food waste, nappies and clothing. In late 2016, John Reed, the founder of Clean up Britain, a pressure group, said: "We need to raise our standards to the levels that the French, Dutch and Germans have reached. These nations are far more enlightened."

That said, around 30% of us are still confused about what can be recycled, according to Recycle Now, a recycling campaign group, and it's hard to believe that this is entirely our own fault. There are also wide variations across the country. In Wales, for example, the recycling rate is 55%, which puts it near the top of the European recycling league (only beaten by Austria), and recycling rates vary hugely between councils. In Newham in East London, it's 15%; in South Oxfordshire, it's 67%. But it is easier to recycle waste if you live in a country house as opposed to an inner-city apartment. In 2012, 10.3 million tonnes of waste were buried in landfill in the UK. This is only 30% of the amount discarded in 1995, and the EU's 2010 target for was comfortably met. In all this, it is particularly important to keep biodegradable waste out of landfill, as this decomposes to produce methane, a potent greenhouse gas, which can leak into the air.

A significant problem with the material that gets recycled is that the quality falls away quickly because it is mixed up with other materials. With newspapers, for example, when they are pulped, the cellulose fibre length gets shorter because of the mechanical process involved, and a certain amount of new wood pulp needs to be added every time so it can be reused to make newsprint of sufficient quality. With many plastics, recycling involves mixing different sorts together, and this muddle limits the uses that the recycled materials can have.

One of the enduring myths of the recycling business is that it keeps materials out of landfill. In fact, this is only delayed. Aluminium is a good example of the problem. The overall recycling rate for aluminium drinks cans in Europe in 2012 was 70%. That was 27.5 billion cans weighing about 390,000 tonnes. The 2020 target is 80%. Although these numbers are large, at this rate of recycling, 90% of the original metal used in can production ends up in landfill after only eight make–use–collect–remake cycles. Albert Bartlett explains the arithmetic behind this in a much-viewed YouTube video.[3]

This inevitable loss of valuable material is one of the ideas driving the circular economy. Dame Ellen MacArthur, whose charitable foundation promotes this approach, calls recycling the "loop of last resort within a circular economy."[4] That is, something only to be done when as much material as possible has been recovered for use. The final irony of recycling is that it needs a waste stream. If there were no waste, there would be no need to recycle anything. So, rather than pestering people to do more recycling (as children are urged to do), perhaps what they should really be calling for is less waste in the first place.

Notes

1 WRAP works with governments, businesses and communities to deliver practical solutions to improve resource efficiency. Web link: wrap.org.uk.

2 WRAP's website carries data and accounts relating to falling carrier bag use in the UK. Web link: ow.ly/x8i6308PTWi.

3 In what he says is the most important video you'll ever see, Albert Bartlet explains the exponential function. Web link: ow.ly/oGaJ308PUbG.

4 *ISBGLOBAL: Circular Economy Easing the Transition.* Web link: isb-global. com/circular-economy-transition.

51

TRANSITION TOWNS

You are most likely to come across the transition movement in the form of a transition town, and probably the most well known of these is Totnes, the small town in Devon, which was the first. It is no coincidence that such developments are all very local in scale, and there are now a lot of them. As the name implies, transition is a movement away from how our economies and societies are currently organised to ones that are more sustainably based. Rethinking our use of money, and how we think about money, is at its heart. Local communities retaining as much money circulating within them is seen by the transition movement as vital to the health of a local economy.

Compare two high streets. The first has a series of shops and businesses: banks, building societies, pharmacists, bread shops, stationers, coffee shops, clothes shops, supermarkets, travel agents, charity shops, dry-cleaners, cobblers, pubs, pound shops, etc., all of which are household names. Most towns have these, which is one reason why small communities can seem the same these days wherever you go. There is a common feature to all these: the surplus money (over and above that needed to pay wages, utility bills and local taxes) flows out of the community to owners, many of which are now multinational companies.

The other high street has all the same services and businesses, and possibly a butchers shop, a fishmonger, and greengrocers instead of supermarkets. And all these are locally owned and run, with a variety of models of ownership and management. Here, money circulates round and round the community, rather than quickly flowing out of it. Because of this, there is a much better chance of developing a diversity of enterprises and businesses that meet local need.

Rob Hopkins, the inspiration behind the transition idea, explains how powerful all this is in a blog, comparing such an economy to a forest ecosystem. He writes:

> If we can protect local economies from the predations of large extractive corporations . . . they will naturally move towards the kind of self-organising, diverse resilient systems that our woodland was modelling. It's a system designed to maximise the wellbeing of its community, through an approach based not on austerity but on being regenerative. It's an economy that isn't about locking resources up, but allowing them to flow at different speeds, to allow some throughput, but to strive to ensure the maximum amount of circulation as possible. It's the thinking that underpins local currencies, although it can happen without them.[1]

Such local currencies are often found in transition communities, and in some cities as well, such as Bristol, Cardiff, Exeter and Oxford. These local pounds are exchangeable, locally, with the national currency and will be accepted as its equivalent by participating shops. Some local councils allow bills for local services to be paid using them. By definition, such local currencies cannot flow out of the local economy, but have to circulate. That said, they might have more symbolic than economic value, as Hopkins implies in his blog.

Transition Network, the charity promoting transition in the UK, says that the movement is seeking to achieve a "low-carbon, socially-just, healthier and happier future, which is more enriching and more gentle on the Earth than the way most of us live today."[2] In its vision, people work together to find ways to live with a lot less reliance on fossil fuels and on the over-exploitation of other planetary resources, with reduced carbon emissions, improved wellbeing for all, and stronger local economies. In this, it is setting out to resist globalisation. From its early beginnings in Kinsale in 2004, and then in Totnes in 2006, the transition idea has spread, and there are now more than 1,100 registered initiatives in over 40 countries. Given that by its nature, transition is at heart a grassroots social movement, an issue it has to grapple with is what its relationships are going to be with local elected officials and government as decision-making, especially about planning and economic development generally, tend to be carried out within those organisations. Whichever way it happens, engagement in local politics by those within transition is essential if the shifts they desire are to be achieved, consolidated and maintained.

Without doubt, the issue about money circulating around a local economy, instead of quickly draining through it, is a powerful one, but the reality is that in any transition community, it does both of these things all the time, and money comes into communities as well by means of the earnings of people who live there, but whose work brings salaries from somewhere else. This is just as well, as none of these transition towns could survive economically without such inflows of capital, or without the ability to borrow money (in the form of a national currency) to invest.

Many who don't live in such communities can see their attractions. They are certainly good to visit, even if buying coffee is a bit of a lottery when you've not got your favourite national or international chain on the high street. Their undoubted attraction to visitors has something to do with being reminded about what high streets used to be like in the past – when good coffee was just impossible, rather than a matter of luck.

Notes

1 Rob Hopkins' forest metaphor is explained in a blog on the Resilience website. Web link: ow.ly/TWLL308KTR7.
2 Transition Network. Web link: transitionnetwork.org.

52

REWILDING

Rewilding is a newly fashionable idea that has come to grip our imaginations in recent years. In scope, it ranges from the very pragmatic to the utterly romantic. At its simplest, it is a practical conservation strategy. One example is the reintroduction of otters into rivers where they used to be common but from which they had vanished because of recent hunting or river management. Another is tree planting in upland areas to slow down the passage of rainwater so that lowland flooding is lessened or prevented. These can be seen as rewilding at its least controversial, unless perhaps you're an angler who regrets the loss of fish, or someone who values treeless moorland.

However, such re-ottering is not seen by everyone as rewilding, even though the otter is obviously a wild animal. This seems to be because we can remember otters in rivers, and their re-establishment is seen as a matter of restoring a recently lost balance (and perhaps harmony) to river systems. Similar success stories can be told of the red kite, the osprey, the pine marten, a range of butterflies, and many more species.

The introduction of the beaver into Devon's rivers is more obviously an example of rewilding, however, although it is very controversial on the ground because this release was not officially sanctioned. The Scottish government has said that it will give the escaped beavers in the Tay and Forth river catchments the full protection of UK and EU law. The only real difference between the reintroduction of the otter and beaver is that of time span, the beaver having vanished from the UK in the 1520s. Thus, re-beavering implies a significant change to those river systems where the animals would now live. It is this change, and its knock-on effects, that make rewilding such a tricky topic.

One conservation argument for the sort of rewilding that beavers represent is that it puts an animal back into an ecosystem where it once played a vital role. Doing this, those in favour say, means that it will do so again. Those against tend to see the downsides of such a strategy outweighing any advantages, with the passage of time and the establishment of different ecological patterns meaning that such rewilding will be disruptive rather than beneficial. Or so the argument goes.

Those really keen on rewilding in the UK don't think about otters, butterflies and trees. They have their eyes on the grey wolf and the Eurasian lynx, each of which used to roam the British Isles almost at will. The wolf disappeared from Britain in 1680, and the lynx in the fifth century CE. Both are regarded as iconic species, and each would now likely find a ready niche within which to live. For example, lynx prey on small mammals and birds such as hares, rabbits, dormice and other rodents, martens, grouse, foxes, wild boar and roe deer. Feasibility studies about reintroducing lynx to Scotland show there is enough prey and habitat to support around 400 animals, and lynx species have recently been successfully reintroduced into parts of Spain.

The Forest of Dean might also be suitable for the lynx, given the very large population of wild boar (genetically mixed with a domesticated pig breed). This idea is both approved of by rewilders and fretted over by others, and so population control by another wild species might just be acceptable to everyone. The lynx has the advantage of being an attractive animal (it has fine pointed ears and a beautiful pelt), which, as we note elsewhere, is helpful in the conservation stakes. It's likely that there will be considerable public sympathy for having the lynx back among us, even if none of us ever get to see it because of its nocturnal, solitary and remote lifestyle. There will also be considerable informed opposition.

The wolf is a different matter altogether. It has been successfully reintroduced into parts of North America (in Yellowstone Park, for example), and is common in continental Europe, even close to Paris. Although also attractive in a rugged, Alsatian sort of way, it comes with a lot of cultural baggage that predisposes people against it. Just think of Little Red Riding Hood, the three little pigs, and all that howling in the night in horror movies. Rightly, or probably wrongly, it would also be seen as a direct threat to humans because of its large teeth, powerful jaws, its pack habit, and an ability to run at around 35 mph. Sheep farmers would certainly need to be philosophical and broad-minded (and subsidised) to welcome it. Again, if the wolf were ever released, it's likely to be in the highlands of Scotland, where the open moors would make ideal territory, and where excess red deer could be taken care of in a more natural sort of way. However, what that might do to the deer stalking economy is something to be weighed in the balance. One thing is probably clear, however: a successful return for the lynx is probably necessary before re-wolfing into the wild could be seriously contemplated. To prepare us for that day, we have Sarah Hall's The Wolf Border and Cormac McCarthy's The Crossing, both of whose compelling novels explore our fraught and contradictory relationship with wolves.[1]

At the romantic end of the rewilding spectrum is the auroch, a huge ox, 6 feet high at the shoulder, large herds of which roamed across Europe in Mesolithic times. They were hunted to local extinction in Britain around 1000 CE and finally became extinct in the mid-seventeenth century, and so are not around to be rewilded. There is talk of a breeding programme (beginning from longhorn cattle) to create animals that look rather like the auroch, although it's hard to see what the point of this is as they would be genetically quite different.

There is, however, more to this rewilding idea than the mere reintroduction of individual species. This is what the activist group Rewilding Britain has to say:

> Imagine our natural habitats growing instead of shrinking. Where space for nature is expanding beyond small pockets of reserves. Imagine species diversifying and thriving, instead of declining. That's rewilding. We could be a country in which bare lands spring back to life and are filled once more with trees and birdsong. We could be surrounded by the thrum of insects, colourful butterflies and moths, wildflowers and fungi.

Rewilding Britain and other groups that support the idea, including writers such as George Monbiot, say that rewilding offers hope not just for wild-life, but for humanity and the planet.[2,3] It's an opportunity, they say, to leave the world in a better state than it is today. Monbiot, in his manifesto for rewilding the world, says that our huge loss of flora and fauna over time is now locked in place by publicly funded conservation policy which says that conservation sites must be maintained in the condition they are now in.[4] He says that this is often a state of extreme depletion that now has to be maintained through intense intervention to prevent change. Monbiot says that rewilding should involve more than reintroducing missing animals and plants; it should involve: "taking down the fences, blocking the drain-age ditches, culling a few particularly invasive exotic species but otherwise standing back. It's about abandoning the Biblical doctrine of dominion which has governed our relationship with the natural world." This is a long way from the return of the otter to its old habitats.

Notes

1 Hall, S. (2015). *The Wolf Border*. New York: Harper Perennial; and McCarthy, C. (1994). *The Crossing*. London: Picador.

2 Rewilding Britain. Web link: rewildingbritain.org.uk.

3 Monbiot, G. (2015). *Feral: Rewilding the Land, Sea and Human Life*. Harmondsworth: Penguin.

4 George Monbiot's manifesto for rewilding the world can be found on his blog. Web link: ow.ly/lo1h3o8KUMX.

53

BIOMIMICRY

The Biomimicry Institute says that biomimicry is "an approach to innovation that seeks sustainable solutions to human challenges by emulating nature's time-tested patterns and strategies."[1] That is, we learn from what nature already does and do it in our own way.

Many everyday examples of this are technological. One is the swimming costume modelled on sharkskin that proved so successful at the 2008 Beijing Olympics; so successful, in fact, that it was subsequently banned from competition. Seen with an electron microscope, sharkskin consists of overlapping small scales that have grooves running lengthways down them. These discourage parasites such as barnacles and make for smooth movement by reducing the drag between skin and water. Coatings have been developed for boat hulls that do the same thing. This mean less fuel is needed and the hulls need cleaning less often, saving time and polluting chemicals. Another is a paint based on the lotus flower's ability to remove dust and dirt particles. The leaf has very small projections that catch debris, and when water rolls over the leaf, it takes anything on the surface with it. The smooth-looking paint actually has a rough surface that repels dust and dirt, lessening the need for washing and cleaning.

But it's Velcro that is probably the most widely known (and used) example. This was developed in the 1940s and mimics the way that seed burrs from the burdock plant can attach themselves to fur and clothing and so get carried away from the parent plant. Velcro has revolutionised how we fasten clothes, shoes and bags, and enabled a new sport in the 1980s, Velcro jumping. In this, people wearing Velcro suits hurl themselves as high up on a wall as they can. Sadly, this is not yet an Olympic sport.

There are other examples, as web searches will illustrate. All these are undoubtedly valuable, but none is likely to make a substantive contribution to what the Biomimicry Institute says is needed: sustainable solutions to human challenges. For these, we may have to look to basic scientific research.

The institute's blog sent a message to the world leaders meeting at COP21 in Paris in December 2015, saying that if you need solutions, you should "ask nature." They then listed "a few of nature's strategies and corresponding innovations that can lead us down a more life-sustaining path."[2] These included:

- developing agricultural systems that mimic the way natural ecosystems function, reducing our agricultural carbon footprint, and creating soils that act as storehouses for atmospheric carbon;
- creating carbon-negative plastics made of methane-based greenhouse gases, inspired by carbon capturing processes in nature;
- copying alpine pine trees whose needles' waxy coating enables them to convert UV radiation into blue light and enhance photosynthesis;
- building on how the saguaro cactus removes carbon dioxide from the atmosphere to form solid calcium carbonate that stays in the soil; and
- transforming carbon dioxide into polymers in a process modelled on the Calvin cycle, which is a key part of photosynthesis in plants.

The last of these is probably the holy grail of biomimicry, as photosynthesis is one of the most important chemical reactions on Earth. Here, plants convert the energy in sunlight into useful chemicals, removing carbon dioxide from the atmosphere in the process and releasing oxygen. It's no surprise, therefore, that there are currently many efforts to replicate it. It's equally unsurprising, perhaps, that this is proving difficult to do with any great efficiency. Not that photosynthesis is particularly efficient, but then it doesn't have to be. Given that every hour, more energy from the Sun falls on the Earth than all the energy we consume in a year, it's not exactly in short supply, and the evolutionary processes that have given rise to photosynthesis as we know it today have evolved to be effective rather than efficient. Indeed, as it is a complex multistep process, it's hard to see how photosynthesis could be more efficient than the 1% it is. Wikipedia (which provides detail on the complex process) says that the average rate of energy capture by photosynthesis across the world is around 130 trillion watts, which is about three times human power consumption.[3] This converts over 100 billion tonnes of carbon into biomass per year in the process, and all this takes place at low temperatures and pressures.

There are two broad stages to photosynthesis in leaves: (i) sunlight breaks up water molecules, producing electrons, hydrogen ions and oxygen; and (ii) the electrons and hydrogen ions combine with carbon dioxide to create carbohydrates such as glucose. The molecule chlorophyll, which gives plants their green colour, is key to the first step, and this is the focus for most of the attempts at mimicry. Sunlight is already used to split water by electrolysis using electricity from solar panels to produce hydrogen that can be stored and used, for example in a fuel cell, to produce power when it's needed. But the process is not particularly efficient or cheap.

As well as the process mentioned above, others are trying to mimic the first part of photosynthesis by using a genetically modified virus, engineered to hold catalysts on its surface to capture light energy. The virus provides a frame around which the components attach themselves at the right distances for reactions to take place that greatly improves the efficiency of the process. As a bonus, the structure actually assembles itself, but there are still problems to be overcome. The prize in all this is huge, as if it can be carried out at scale and cheaply, it means that the carbon-neutral fuels needed for transport, heating and other essentials can be made and no additional CO_2 gets added to the atmosphere. This means that fossil fuels can be left in the ground along with the carbon dioxide they might have produced. This is the holy grail, unless you're an oil or coal producer, that is.

Notes

1 The Biomimicry Institute has a blog saying to world leaders: "If you need solutions, ask nature." Web link: ow.ly/rfDv3o8PTfY.

2 The Ask Nature website explains how polycarbonate polymers can be made from carbon dioxide. Web link: ow.ly/MU82308PSQk.

3 Wikipedia's page on photosynthesis is a starting point for exploring this complex subject. Web link: ow.ly/IJQT308PTtn.

54

(ENVIRONMENTAL) EDUCATION

In the beginning was the biosphere, and well before our development of agriculture, all education was environmental, because it was a matter of survival: a response to questions of food, shelter and safety. In our success in learning how to understand and manipulate the environment – to read it and to write it – humans have become steadily detached from the world that supports us, with our lifestyles being increasingly technology-mediated and -dependent. While human ingenuity has resulted in more people being fed than Thomas Malthus dared to dream of, we now have a long list of concerns that we can (if we wish) worry ourselves sick about. As we have seen, these include loss of forests, habitat destruction and a steady reduction of biodiversity and species, stratospheric ozone depletion, wide-ranging pollution of air, land, rivers and oceans, desertification, unstable weather patterns, and potentially catastrophic climate change.

Rachel Carson's 1960s book *Silent Spring* is popularly regarded as the point when we became aware of the negative impacts of our industrialised society.[1] Carson's main focus was the indiscriminate use of chemical pesticides, and their residues in food, but, in quoting French biologist Jean Rostand,

"The obligation to endure gives us the right to know," Carson also saw the link between education and our survival.

Others, however, point to much earlier examples of environmental education: for example, social movements in nineteenth-century European industrial cities gave rise to legislation to counter damaging air and water pollution and disease. The educational focus of such movements was not directed towards children in schools, as these were thin on the ground, but towards politicians and public opinion. Over time, the main strands in the development of environmental education have been nature study, outdoor education, rural studies and conservation education, and well before Carson's alert to the wider public, these different paths and emphases were developed across many countries. By 1970, the recognition of the need for an educational response to pressing environmental and social challenges prompted the promotion of environmental education as a significant idea. The World Conservation Union described it like this: "A process of recognising values and classifying concepts in order to develop skills and attitudes necessary to understand and appreciate the inter-relatedness among man, his culture and his biophysical surroundings."[2] It is clear from this that environmental education is not just about raising awareness of issues. It seeks to develop an understanding of the relationships between human culture and our life support system, and emphasises environmental responsibility through social action and personal behaviour.

At the end of the 1970s, the goal of environmental education was summed up by Bill Stapp and his colleagues in this way: to foster an "environmentally literate global citizenry" that will work together to build an acceptable quality of life for all people.[3] These days, environmental education is a broad church whose members have wide-ranging interests and concerns, but all encourage the use of the environment to learn about the human condition and about our relationship with nature.

The range of interests and goals is huge: from those interested in the study of the processes of nature in order to understand them, to those interested in sharing the joy and fulfilment derived from nature in order to bring about significant life-enhancing and life-changing experience for learners, across to those promoting nature as a metaphor for a preferred social order – which as we have seen may be cooperative or competitive, according to how you prefer to see things. Such learning takes place in schools and local communities, with carefully designed interventions, or results from personal or incidental learning with no instructor or guide in sight. As Wordsworth suggested some 200 years ago, you can "let nature be your teacher."[4]

These days, a wide range of charities and other groups both encourage environmental education and provide leadership. By and large, governments do some of the former but little of the latter. This is largely because across cultures, the dominant education model remains focused almost entirely on a productive, competitive economy, and reflects our industrial era, engineered with linear thinking modelled on machine metaphors. This reflects a model of mass production, complete with its attendant quality control and rejection mechanisms. As Fritz Schumacher put it, "If still more education is to save us, it would have to be education of a different kind."[5] However, it is not obvious that environmental education in its current form is this different kind of education, and anyway, its influence remains at the margins of young people's experiences in school. It generally owes any impact more to teacher enthusiasm than to school leadership, policy or values.

Because school education as we know it has helped human society rapidly to develop in an unsustainable manner, it must be open to doubt that it can be changed sufficiently to reverse this trend. Moreover, it is hard to see how a formal school system could ever be in the vanguard of such social change given that schools are a reflection of current social and economic ideas rather than the other way round. This is a classic double bind: in order for schools to change society in such ways, they would need to be changed by society first. Nowhere are there convincing cases of this happening.

Although education systems tend to focus on narrow sets of readily examinable outcomes, work with very young children can emphasise the development of the whole person and a deeper understanding of the world. England's early years foundation stage does this across three prime areas: communication and language; physical development; and personal, social and emotional development. It is after this initial stage that the curriculum narrows. Although extensive work was done by the Cambridge Primary Review to build on that broad foundation stage, its proposals have been rejected by successive governments.

Elsewhere, there are examples of a more connected world view being encouraged through a curriculum that has evolved out of a combination of local issues, social investigation and national priorities. For example, this is the case in Tanzania, where a locally negotiated curriculum involving parental priorities forms the basis of Complementary Basic Education in Tanzania (COBET). COBET serves learners who cannot access the state system, yet achievement levels among children following the programme for one year have been reported to outstrip their counterparts who have studied in the state system for two years.

It is clear from COBET (and other such developments) that the way we teach is important, and there are many examples of effective approaches that have involved pupils working outdoors and learning how to take action to address local issues. Such change-oriented projects are not well established in the UK school system, but there is a long tradition of this action competence approach in Scandinavia. Here, the word action has a specific meaning. It must be focused on bringing about change and be initiated by the students themselves. In this way, the teacher becomes a support to students seeking information from a variety of sources within their community. Crucially, however, it's not any change that occurs that matters, but what the young people learn and the skills and attitudes they develop.

Of course, it's not just schools and universities that function as centres of learning. This also applies to the wider society, where its businesses, shops, farms, media, factories, courts and offices are places where people continually learn. Such locations are in their turn legitimate places for young people to learn through change-oriented projects. In this way, we might expand the idea of environmental education in order to nurture an educational environment. In time, a connected, community-based approach may yet emerge and help us to build the sort of relationship with nature that we need. It might still be a long way off, but one day we might achieve what WWF's 2016 *Living Planet Report* said we needed: "A shared understanding of the link between humanity and nature [that] could induce a profound change that will allow all life to thrive in the Anthropocene."[6]

Notes

1 Carson, R. (1962). *Silent Spring*. Cambridge, MA: Riverside Press. Web link: ow.ly/GMXq308PSem.

2 IUCN (1970). *International Working Meeting on EE in the School Curriculum. Final Report*. Gland, Switzerland: IUCN, UNEP and WWF.

3 Stapp, W. et al. (1979). Towards [a] National Strategy for Environmental Education. In: A.B. Sacks & C.B. Davis (eds), *Current Issues in EE and Environmental Studies*. Columbus, OH: ERIC/SMEAC.

4 William Wordsworth's poem "The Tables Turned" can be found on the Poetry Foundation's website. Web link: poetryfoundation.org. Type "tables turned" into the search box.

5 Schumacher, E.F. (1973). *Small is Beautiful: A Study of Economics as if People Mattered*. London: Random House.

6 WWF (2016). Living Planet Report. Web link: ow.ly/rIEv308PSCc.

55

THE MONTREAL PROTOCOL

We end the book with a good story about effective international action to address a serious global problem. As we saw in the chapter on ozone, it was in the early 1970s that James Lovelock, who developed the idea of Gaia, first detected chlorofluorocarbon (CFC) gases in the atmosphere. By 1974, a prediction had been made that CFCs would decompose, release chlorine into the stratosphere and damage the ozone layer, and in 1977 the United Nations Environmental Programme (UNEP) held the first international meeting to address the problem. Although the US banned all non-essential use of CFCs in aerosols in early 1978, their use continued to increase across the globe. By 1982, low ozone concentrations had been found over Antarctica, and in 1984 the British Antarctic Survey found that there had been a thinning of the ozone layer. This discovery led to swift international action.

March 1985 saw the Vienna Convention for the Protection of the Ozone Layer, and in September 1987 the first Montreal Protocol on Substances That Deplete the Ozone Layer was signed. In the following year, CFC production began to be phased out, and was ended completely in 1996 in the US and Europe, by which time the evidence for CFCs damaging the ozone layer was overwhelming. Although global CFC production only finally ended in 2010, its concentration in the atmosphere had been falling since 2000.

All this was probably a close-run thing, and it will take a very long time to remove all the CFCs completely from the atmosphere because they are so unreactive. Their story illustrates a number of points.

1   Achieving the Vienna Convention and the Montreal Protocol in such a short time was a great achievement by UN agencies.
2   This contrasts with the huge mount of time it has taken to make progress in relation to reducing climate change.
3   CFCs were produced in good faith by ethical scientists, and used to better the lives of millions of people over a long period.
4   International scientific collaboration was effective at identifying the problem, and providing the understanding to sort it out, but it still needed political will and action to make it work.
5   The protocol won support from chemical companies that made money producing CFC substitutes that did not damage the ozone layer – HFCs.
6   It won the backing of developing countries because alternatives existed, and they were given money to ease the transition.
7   Although it was known that the replacement HFCs were greenhouse gases, and therefore also a problem, it was worth making the switch as it was so important to replace CFCs and protect the ozone layer.

As we said, this is a good story, which is even better than has been presented so far because it's now clear that the Montreal Protocol has been our most potent strategy against climate change, even though it wasn't designed to be that. The Economist reported in 2014 that the protocol, by preventing the continuing release of CFCs (which were powerful greenhouse gases), had stopped the release of the equivalent of 135 billion tonnes of CO_2. The Economist also said that if we could replace the HFCs as well, this would be like removing another 130 billion tonnes.[1]

Happily, in October 2016, the UN concluded a deal in Kigali to phase out the use of these chemicals, starting in 2019 with economically developed countries cutting use by 10% (and by 85% by 2036).

By 2024, more than 100 developing countries, including China, are set to join in. A few other countries, including India, Pakistan and some Gulf states, insisted on waiting until 2028, saying that their economies needed more time to grow. The politics of climate change are such that this had to be agreed.

It is said that when all these changes work through, there may be enough greenhouse gas removed from the atmosphere to be the equivalent of up to 0.5°C of global warming, thus buying us more time to tackle the main problem, CO_2 emissions. The scale of this challenge is huge as HFCs are currently used in every fridge, freezer and air conditioning unit, and in many aerosols, and many of these systems will need to be changed to take the hydrocarbon replacement gases. Some worry that the delay in implementation may cost us dear. Others say that the changes will happen much faster than we imagine because once industrial production of the replacement gases and machines starts, it will rapidly spread across the world. As ever, time will tell.

The Montreal Protocol was successful because the serious and urgent nature of the problem was acknowledged, and because of the partnership and trust that was established between science, the chemical industry, and politicians from both economically developing and developed countries. Everyone had a positive role to play, and a lot to gain by doing so.

As we noted in our opening chapter, the 2015 Paris Agreement has set a target to limit global temperature rises to no more than 1.5°C above pre-industrial levels, but it is a sobering thought that the partnership and trust that was at the heart of the Montreal Protocol process is missing (so far) from the Paris Agreement.[2] The science is still disputed by a lot of groups, the agreement is not legally binding, and there is a clear reluctance by richer countries to pay the price that is needed. As we note above, the agreement requires a transfer of $100 billion a year from economically developed countries to developing ones by 2020. Whether this will happen is anyone's guess, as is how effectively any money will be spent.

However, the existence of the agreement remains a very positive thing, and a majority of countries have now signed up to it. Although much still needs to be negotiated, the agreement marks a significant shift away from an age where economic development was powered by fossil fuels to one where renewable energy will hold sway. The challenge is to make that a brighter and cleaner future for everyone.[3]

Notes

1 *The Economist* (2014). The deepest cuts. 20 September. Web link: ow.ly/9P8r3og3Jkk.
2 The UN Climate Change Newsroom has a range of articles and agreements about climate change, including the Paris Agreement. Web link: unfccc.int.
3 UNEP's Ozone Secretariat has an online handbook of the Montreal Protocol. Web link: ow.ly/3kMv3o8QKVA.

INDEX

Taylor & Francis eBooks

Helping you to choose the right eBooks for your Library

Add Routledge titles to your library's digital collection today. Taylor and Francis ebooks contains over 50,000 titles in the Humanities, Social Sciences, Behavioural Sciences, Built Environment and Law.

Choose from a range of subject packages or create your own!

Benefits for you

- » Free MARC records
- » COUNTER-compliant usage statistics
- » Flexible purchase and pricing options
- » All titles DRM-free.

REQUEST YOUR **FREE** INSTITUTIONAL TRIAL TODAY

Free Trials Available
We offer free trials to qualifying academic, corporate and government customers.

Benefits for your user

- » Off-site, anytime access via Athens or referring URL
- » Print or copy pages or chapters
- » Full content search
- » Bookmark, highlight and annotate text
- » Access to thousands of pages of quality research at the click of a button.

eCollections – Choose from over 30 subject eCollections, including:

Archaeology	Language Learning
Architecture	Law
Asian Studies	Literature
Business & Management	Media & Communication
Classical Studies	Middle East Studies
Construction	Music
Creative & Media Arts	Philosophy
Criminology & Criminal Justice	Planning
Economics	Politics
Education	Psychology & Mental Health
Energy	Religion
Engineering	Security
English Language & Linguistics	Social Work
Environment & Sustainability	Sociology
Geography	Sport
Health Studies	Theatre & Performance
History	Tourism, Hospitality & Events

For more information, pricing enquiries or to order a free trial, please contact your local sales team:
www.tandfebooks.com/page/sales

The home of
Routledge books

www.tandfebooks.com